A Knee and Shoulder Handbook For All of Us

I hurt My Knee.
Oh, it's my Shoulder.
What do I do now?

By Alan M. Reznik, MD and Jane Y. Reznik

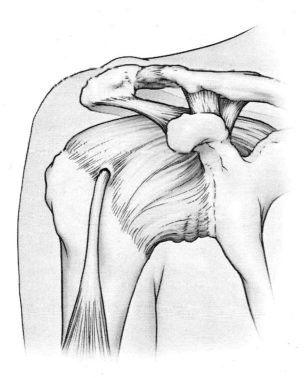

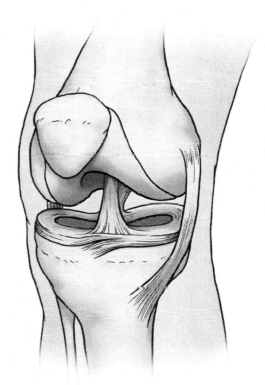

For the Best of Health,
Alan M R MD

FIRST EDITION

ISBN #: 798-557-64071-3

NOTICE

This book is dedicated to the memory of my father, Barry D. Reznik, as an engineer, an inventor and a truly self-made man. He believed that almost anything could be explained to anyone if you had a little time, a napkin and a pen. His presence is deeply missed.

AMR

Table of Contents

Acknowledgment

Where does one start when thanking the people who contributed to this project? You could go back to birth and thank your parents, or stop short of that event and thank the first teacher that thought you might have potential. You may want to thank your wife for lovingly putting up with the trials that ensue once anyone starts a book of any type. These are the people who believed in me and told me to never give up, and I thank all of them. I have also grown to appreciate many others, including the clinical faculty at New York's Mount Sinai Hospital during my orthopaedic residency, Professor Robert Duthie and everyone on the "top floor" at Oxford University's Orthopaedic Department, Dale Daniel, Ray Sachs, Mary Lou Stone and Don Fithian from the San Diego sports fellowship. I would also like to acknowledge the Yale Orthopaedic Department's faculty, staff and residents and my current partners at The Orthopaedic Group in New Haven, CT. All of these people have taught me everything about orthopaedic surgery and helped to shape my daily clinical practice.

I have to especially thank a former Yale orthopaedic resident, Robert Edward "Ted" Kenyon, MD, for telling me that the patient booklets I had written for my office inspired him to write his book. I would also like to thank him for returning the favor by inspiring me to write this one. I owe many thanks to my wife Elizabeth who, in her role as a photographer, has contributed many of the photos presented here and my daughter Michele, who is a communications designer, for the cover. Finally, with great pride, I am grateful to my daughter, Jane, for at age 18 thinking it wasn't a terrible idea to work with her dad on a project that may help patients better understand what is wrong with their shoulder or knee and hopefully help them to a better diagnosis and a speedy recovery.

Photo and Art Credits
Laura Kovalcin - Anatomy drawings of knee and shoulder
Natasha Peavy - Back cover photo of Jane Reznik
Elizabeth Reznik - Surgical and therapy photos
Michele Reznik - Front and back cover design
William Schreiber, MD - Sports photo
Aspaeris.com (ACL prevention shorts) - Valgus collapse diagram

Doctor's Introduction

Remember the Peanuts specials during the holiday season? Remember how it sounded when the adults spoke to Charlie Brown? To me, that is a fairly accurate picture of what it's like to listen to a doctor after he says the words, "looks like you may need surgery." I could see it in my patient's eyes. Just like in Charlie Brown, from that point on, all they heard was "wah-wah wah-wah…" Telling anyone that surgery was an option seemed to cause both short-term memory loss and retrograde amnesia. It was worrisome. No one was listening to a thing I said. With this in mind, I started to write blurbs to give to my patients as they left the office. The "wah-wahs" sounded clearer when written down. A piece of paper with a summary of the problem they had, the possible treatment and the care afterwards was a great help. Like homework, they were to read it and write down their questions for the next visit. They were even encouraged to bring family with them to go over the answers.

In time, the papers grew into little booklets with pictures. Still, it was clear that patients had a limited understanding of what was wrong. They could not see that in using an arthroscope in surgery, dozens of different operations can be done through the same small incisions. Patients who had completely different operations would ask why it was taking them longer to heal. So I posted a few YouTube videos for my patients. They could now see the different procedures on video (www.youtube.com/DrAReznik). My nurse and I also wrote a list of frequently asked questions (with the answers) and even an article on how to make the most of your office visit. All of these educational pieces seemed to help in a small way. However, as certain issues were solved, other issues crept in. The common complaints of knee locking, buckling and giving way were poorly reported, so I wrote an article on that. Lastly, too many children were being injured in sports. Most injuries occurred because of a lack of understanding of simple safety precautions for growing children or a lack of appreciation of what special concerns we have for injuries to growing bones. To help, a newsletter called "Playing it Safe," based on the American Academy of Orthopedic Surgeon's Play it Safe program, was sent out to the local pediatricians. Soon enough, it became clear that others might benefit from the collection of patient materials, but mailing them to everyone was impractical. Thus, this book was born.

Inside, you will find the articles described above, as well as articles on a series of common knee and shoulder problems. I have presented what I believe are the latest treatment concepts, tips on understanding the diagnoses and preferred treatment options. Of course, like any book on medical conditions, this cannot substitute for an exam by an experienced physician or make a diagnosis, but once a diagnosis is made, the information here may help make it easier for you to have a good understanding of your problem. In my experience, the most well educated patients can understand the reasons for their rehabilitation program and therefore participate fully in their own recovery. It is my sincere hope that this understanding, along with expert treatment by your own doctor, will improve your chances of an excellent outcome.

Alan M Reznik, MD

Injuries in Children

Play It Safe!

In the U.S. alone, there are over three quarter million ER visits each year by children under the age of 15. A major cause of this is the alarming rate of injury during sports. The "Play It Safe" program was created by the American Academy of Orthopedic Surgeons (AAOS) to increase awareness and to reduce injury to children during these sporting activities. Many studies show that the majority of injuries occur in unorganized or casual sports, like pick-up games of basketball, baseball and football. Still significant, organized league sports make up about one third of the injuries. Reducing the risk of both injury rates is a goal of the "Play it Safe" campaign. There are four topics that require special attention when it comes to children and sports:

1- *Young athletes are not just small adults*
2- *Growing children's injuries create special concerns*
3- *Diagnosis and treatment offered to children*
4- *Prevention*

Young athletes are not just small adults

Children are growing all the time. This gives them some special advantages over adults. To start, their bones have a little more spring and bend before they break. They are typically lower to the ground and have lower body weights, making most minor falls of little consequence. At the same time, they tend to be less prepared for injury, and their sense of danger is far less than an adult's. Children also grow at differing rates at different times during development. A sudden growth spurt or a change in limb length can create the gawky behavior that makes some children seem accident-prone. These factors alone help explain some of the injuries prevalent in child athletes.

Sports injuries I've seen in children vary from a child simply exceeding his or her physical limitations to an accident occurring during an unsupervised activity in an unsafe environment. Adults have to be aware of their own children's limits. Some 14 year olds are fully-grown, while others are not. I often hear stories of children in an age-based league playing against kids who weigh 50 to 100 pounds more than them. Many coaches and

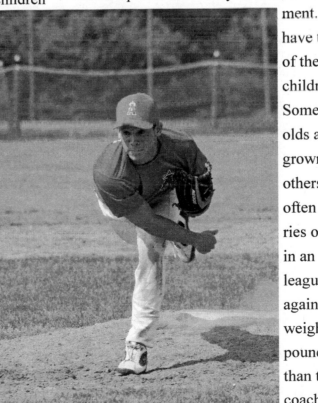

Photo courtesy of Dr. William Schreiber

parents take performance at very young ages to an extreme, and lying about a child's age or weight for an unfair advantage is simply wrong. Worse yet, studies have shown that in the late 1990's, up to 500,000 young athletes were using black market steroids to increase muscle mass. The risks of these drugs are widely known, serious for children and potentially life threatening. These "performance enhancing" products should be completely avoided. Sports for children should always be fun, not unhealthy or dangerous.

Growing children's injuries create special concerns

The bones in children grow in specialized areas near each joint called growth plates. These areas are softer than calcified bone in the middle of the limbs and therefore are more susceptible to injury. When an injury to a growth plate occurs, future growth and alignment of the limb are at risk if it is not treated properly.

Growing children's bones can buckle and bend without breaking all the way. This creates fractures in the middle of the bone, which are often known as "green stick" fractures since they resemble what happens when you try to break a growing tree branch. These green stick fractures break, deform and stay deformed, even though part of the bone (branch) is still intact. These "incomplete" fractures or breaks also require special attention. Both growth plate and green stick fractures affect bone growth. The growing child can remodel broken bones and overcome minor disturbances in growth. The child's age, the fracture location, the bone's

angulation (amount of bend) and the fragment's displacement (separation) are important factors in determining how well a fracture will heal without orthopedic intervention. When it is clear that a deformity will result, prompt treatment is necessary.

Diagnosis and treatment offered to children

Occasionally, because the growth plates are not calcified, a diagnosis can be much more difficult to make. The cartilage and growth plates cannot be seen directly on normal X-rays. In these cases, a precise history of the injury (force involved, position of the limb, direction of impact and anatomic location of the injury) helps in making the correct diagnosis. Many times, there is no substitute for an examination by experienced hands because findings on an x-ray are often only seen once the healing process is well underway. Sometimes a missed diagnosis becomes a missed opportunity for a simpler treatment.

In children, some injuries can be treated with a sling, splint or cast while others require perfect reduction to realign the growth plate or the joint space. The goal of all treatment should be to use the least invasive method to allow a child's bones to heal with the lowest risk of future deformity or loss of future growth.

Prevention

Young athletes should be encouraged to play in organized or supervised sports. They should have training or specific stretching and exercise programs to prepare them for the sport. The sport itself with or without preparation should not be their only exercise. A child's coach and parents should take into account the child's age, height and weight before matching them in age-only based sports. Parents should be sure the coaches have appropriate training and qualifications to coach their children. The children must have access to a safe playing area and appropriate, well-maintained equipment. Field conditions, weather conditions and available supervision should always be a factor when deciding to have a competition. In hot weather, parents should be sure their children are well hydrated and be aware of the risks of hyperthermia on very hot and humid days.

Children should warm up and stretch before participating in sports. They should drink plenty of fluids. They should have appropriate fitting equipment. They should tell the supervising adult when they are hurt instead of trying to play through painful injuries. Safety rules for growing athletes, such as pitch counts, should be strictly followed, and children should not be played in multiple leagues in the same sport in the same season to "get around" these rules, no matter how great the parents think they are. The rules are designed to protect growing children from injury; ignoring the rules will risk serious growth injury and only shorten their playing careers. Children should never be given "performance enhancing" drugs or supplements.

Protective gear is also important and sports specific. Helmets for biking, skiing and roller-skating are no longer optional. Mouth guards, shin guards and plastic face guards have helped to reduce injuries and should be used. Elbow pads and wrist protectors should be worn for inline skating, even on pathways designed for skating. Binding releases for skis should be calibrated to the child's skill level, height and weight each season.

The AAOS promotes the idea that "youth sports should always be fun. The 'winning at all costs' attitude of coaches, parents, professional athletes and peers can lead to injury." Remember, having unrealistic expectations can lead a child to continue play despite warning signs of injury. This puts a child at increased risk. Lastly, the AAOS reminds us, "Coaches and parents can prevent injuries by fostering an atmosphere of healthy competition that emphasizes self-reliance, confidence, cooperation and a positive self-image."

Pitching limits for growing children and young adults

We have all heard the expressions: "His arm is burned out." and "Be careful, or you might throw your arm out." But what is this all about? In a young athlete, a "thrown out arm" could be a result of overthrowing due to an overzealous coach, parent or even the athlete him or herself trying to get around pitch counts. Sadly, this practice is wide spread. After all, who doesn't want to win a tough game? Everyone thinks, "Why not use your best pitcher in a 'pinch?'" "How can it hurt?"

Yet it can hurt. All young athletes are growing, and there are very delicate special areas of the bone that provide this growth. Immature cartilage expands and swells, and these cells are eventually replaced by new bone. This special area of bone is called the growth plate. It is vulnerable to stress and repetitive motion. If damaged, growth can be stopped or stunted. The soft cartilage around the growth plate can also crack and fail causing loose bodies (like OCD in the knee. See the chapter on loose bodies in the knee). These injuries can ruin even a bright pitching future and, in most of these cases, overthrowing or ignoring the pitch count is the major culprit.

What are some basic pitching rules?

No curve balls before age fourteen.
For younger players, no more that 5.5 pitches per year of age per game (a good rule of thumb).
In young players, no more than 1000 per full season.
For older players (with more mature bones), no more than 75 pitches per game
and, no more that 2000 pitches per full season.
No more than 8 months of pitching per year.
No dual seasons (pitching for more than one team in the same season, this is much more common than one may expect).

So with these rules, why are kids still getting hurt? Everyone likes to believe that his or her kid will become the next Tom Seaver, Nolan Ryan or Roger Clemens. The coaches know this and are taking their cue from the excited parents who truly believe his or her is kid is superman (or some other invincible comic character). The result is that they all want to illegally get around the rules, even when the leagues are highly supervised. So coaches and parents let them pitch all year round, "forget" the date of the last full game they pitched or worse, they neglect to press the pitch counter button in the middle of the game so the count can be "accidently" extended (this last one I would not believe if I had not seen it myself during a heated game between two local rival teams).

So what can we tell those parents and coaches? They need to know that the average age that most professional pitchers start pitching is 17.3 years old and many don't even pitch before the end of high school or early in college. Over-pitching a promising ten year old has never resulted in a major-league star. Knowing this, dual seasons should be outlawed. A season off allows recovery of the growing areas of bone. In more mature players, above age 14, light lifting in season helps strengthen other muscles and prevent some injuries. But remember, heavy lifting can stunt growth and creates more slow twitch fibers as opposed to fast twitch muscle, which will decrease throwing speed. You can overdo lifting! Core and balance exercises are key to good form and help transfer force from the ground to the ball. This takes stress off the arm, elbow and shoulder (Tom Seaver's legs were said to be like tree trunks). If form is an issue, have a pitching coach check it and make the needed corrections. Always pull pitchers if a pitch hits the ground in front of the plate or they throw over the backstop. These are certain signs the star pitcher is getting tired and an injury is right around the corner.

Knee Pain in Young Children and Teenagers

Pain in the front of the knee is one of the most common problems seen in young children and adolescents. The symptoms of possible knee injuries range from constant pain to pain only with heavy activity. The potential causes range from tendonitis to a growth plate injury. In this chapter, we will look at a number of these problems and their treatment.

Osgood Schlatters disease

This is one of the more common causes of anterior knee pain in a growing child. It is likely to occur during sports that include running and jumping. It is most common in basketball because of the fast starts, quick stops and frequent jumps. These repetitive motions put weight on the kneecap (patella), the tendon attached to it (patella tendon) and its attachment to the tibia (the tibial tubercle). In children, the tibial tubercle is directly above a part of growing bone called a "growth plate". The constant traction provided by some sports on this area can injure the growth plate under the tubercle. Repetitive injury causes the growth plate to stretch or enlarge, and the bone at the tendon attachment enlarges as well. When the growth plate is injured, the enlarged bone can be tender, and the plate itself can hurt when weight is put on it. This explains why going up or down stairs, kneeling, running or jumping can increase knee pain. The problem is caused by microscopic fractures in the growth plate. Fur-

thermore, and with enough force, the tubercle can break on rare occasions. Sometimes a small part can break off, causing a separate piece of bone to form within the patella tendon. This can also be painful.

Treatment for Osgood Schlatters (OS) disease

In general, early treatment for this problem is based on the level of symptoms present. If there is no bump, and it is only mildly tender over the tubercle, rest, ice and NSAIDs may be enough treatment. The patient can safely return to sports when his or her symptoms are gone. While the area is tender, the youth should avoid all activities that cause pain. This includes squatting, kneeling, running, stair climbing (particularly as an exercise) and jumping. The adage, "no pain, no gain," does not apply here. In more moderated cases, there is a bump and a gap or a widening of the growth plate that can be seen on an X-ray. In these cases, the patient may benefit from immobilization with a knee immobilizer, then a rest period and a slow return to sports. In severe cases, when ambulation is painful, X-rays should be reviewed to be sure that there is no fracture. The exam should focus on the integrity of the patella tendon's attachment to the bone. If this is compromised, it may need to be fixed, although repair runs the risk of an early closure of the anterior growth plate and a slight risk of a back bowing of the tibia (recurvatum). Early treatment in all cases is best, and the patient's vitamin and calcium intake should be reviewed as part of the history. Supplements should be given if his or her intake is poor, in order to help with the healing process. Chronic, recurrent cases of OS may require casting to rest the tissues and allow them to heal. In some of the cases in which it is a chronic problem, a small piece of bone may have broken off and embedded in the tendon. Later, and even as an adult, this fragment may become even more problematic. Removal of this fragment is occasionally needed.

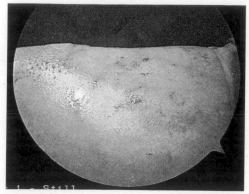

The knee after removal of the bone fragment.

Jumper's knee

This is a similar problem occurring on the patella side, as opposed to the tibial tubercle side, of the knee. In this case, the traction causes the lining on the patella (its growth plate) to pull off the bone, causing pain and weakness. The traction causes new bone to form, just as it does in Osgood Schlatters disease, only now on the tip of the knee cap. It is treated according to the symptoms as explained above. In rare cases, it can also be associated with a fracture, called a "sleeve fracture". This is when a small fragment of bone with a sleeve of periosteum (membrane that lines the outer surface of the bone) pulls off the kneecap, weakening the attachment of the tendon and making use of the knee impossible. Operative repair is needed in severely displaced cases.

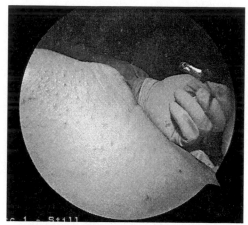

Osgood Schlatters lesion pre-op.

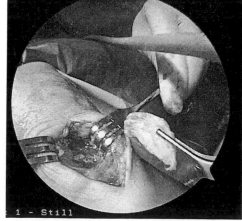

Removal of an Osgood Schlatters lesion.

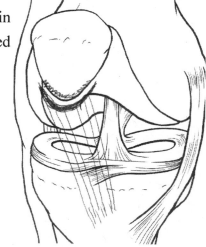

Patella tendonitis

This is the lesser of anterior knee pain problems. In this problem, the tendon that connects the tibial tubercle and the patella is inflamed. Once again, rest, ice and NSAIDs may be enough for treatment, and patients can safely return to sports when their symptoms are gone.

Growing pains

This is the term used by many to explain any knee pain in children. I believe in "growing pains" as its own diagnosis since rapid bone growth can cause pain as the soft tissues, tendons and ligaments try to catch up with it, particularly during a growth spurt. However, this diagnosis should only be used after a complete exam, including X-rays yields no concrete findings, like those mentioned in this chapter. NSAIDs, ice, rest and gentle stretching can help ease the pain.

Kneecap or patella femoral pain is discussed in the "Kneecap Pain and Dislocations" chapter.

Osteochondritis Dissecans

In Osteochondritis Dissecans (OCD), a fragment of bone below the joint surface loses its blood supply and, along with the cartilage covering it, separates from the rest of the bone. The bone fragment with its cartilage cover is referred to as the osteochondritis lesion. Without blood flowing, the bone under the surface is not viable. The synovial fluid of the joint can continue to nourish a broken fragment, and it may enlarge. The fragment can then cause locking of the joint, pain and swelling. Treatment of this problem depends on several factors: the age of the patient, the size of the lesion, its location, the condition of the bone base it came from and the extent it is still attached to the bone.

The most common place for OCD is the knee. The bottom of the thighbone (femur) is made up of two curved surfaces next to each other with a notch for the ACL and PCL in between. These surfaces are called the medial and lateral femoral condyles. The medial con-

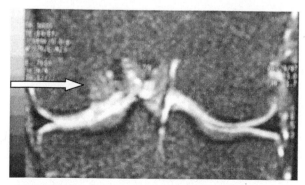

OCD lesion in the lateral aspect of the medial femoral condyle, The most commonm location in the knee for OCD.

dyle is on the inner side of the knee, and the lateral is on the outer side. The condyles are covered with articular cartilage to ensure the

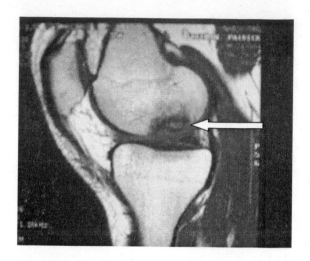

Side view on MRI of loose body from an OCD lesion

smooth movement of the joint. OCD can occur in the lateral condyle or on the patella. However, the most common place in the knee is the lateral side of the medial femoral condyle, more towards the middle of the joint (see MRI lower left and above, the arrow points to the defect). This is because the inside of the knee bears more weight. The involved area is near a raised spot on the tibia called the tibial eminence. It is believed that loading and twisting causes the tibial eminence and femur to hit each other and microscopic fracturing to occur. If the injured area is under constant stress from weight bearing, running or repetitive trauma, it won't have time to heal. The blood supply to the area fails, and the fragment loosens.

Still, even with a good theory, the exact

reason for this disease and the associated bone breakdown in any specific individual is often unknown. Most doctors believe that it is due to repetitive trauma, but an underlying vascular problem in the local bone under the lesion may have a role. It could also be from an unnoticed injury, steroid use, elevated blood fats or a genetic predisposition. It occurs commonly in older children or adolescents who participate in sports. Doctors also believe that the repetitive motion in sports can cause a small segment of the bone to fatigue and fracture under the surface. If there is continued micro-trauma from repetitive loading (for example, running on an already injured knee), it will prevent the defect from healing and again, loosen the bone fragment. This loose fragment causes swelling and pain. If the fragment comes off completely, it's called a loose body. Loose bodies could get trapped in between the joint and cause locking of the knee. The defect, without the cartilage present, is not smooth, so the bones rubbing against each other could cause arthritis over time.

How will I know I have OCD?

If you have OCD, you will feel pain in the knee, especially after being active, and have some swelling. You may experience catching or locking and difficulty with full extension of the leg. If the fragment breaks free or enlarges, symptoms will only get worse over time. Eventually, it may become too painful to put any weight on the leg at all.

OCD is usually diagnosed with X-rays (including a special notch view). To stage a

lesion and see if there is any bone damage or if there is fluid under the lesion itself, an MRI may be ordered. If there is fluid present, the cartilage is loose and surgery is required. Sometimes OCD is recognized when you are being tested for something else. In growing children, an early lesion that is not loose could heal with crutches, rest and a brace or cast. Again, in children as in adults, if the defect is displaced or loose, or if an MRI shows fluid under it, surgery may be necessary.

Treatment options

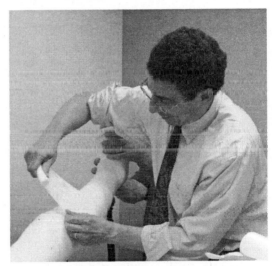

Dr. Reznik applying cast padding before appling a long leg cast

If the bone and cartilage aren't completely detached from the bone, nonsurgical treatment may be appropriate. The younger you are, the more likely it will be successful. It may require up to six weeks of immobilization and using crutches so you don't put any weight on the injured leg. It could take three months to heal so a brace can also be helpful. After casting you should avoid any activities that cause pain. Remember, after long periods of immobilization, a good physical therapy program to build

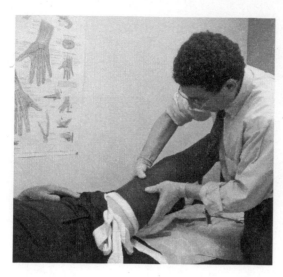

Dr. Reznik applying a fiberglas cast.

the muscles around the knee is always important.

If conservative treatment is unsuccessful, the fragment is unstable or in an older patient, surgery is the best choice to restore the smooth surface. There are several options for surgery, depending on the state of the fragments. First, if the cartilage hasn't broken loose, it can be drilled to stimulate a new blood supply, and it can be fixed in place using pins or screws that are sunk into the cartilage to hold it in place while it heals. If biodegradable pins are used, they do not need to be taken out. Sometimes the damaged fragment may not fit perfectly back into place. The bone around the defect may have also changed, and the surgeon will therefore need to re-contour it, rebuild it or graft it. If the fragments are completely loose, fragmented or missing altogether, the surgeon might clear the cavity to reach fresh, healthy bone and attach a bone graft into position with screws or pins. If it is loose and can be replace the peice should be put back and fixed in place.

When there is complete bone loss, the bone may have to be replaced. It will usually be with a local autograft, an allograft (from a donor) or a scaffold material. An autograft (from your own body) works well, but there is concern about loss of cartilage at the donor site. The doctor will try to pick a non-weight bearing donor site that won't cause any pain or further problems. Finally, fragments from smaller shallow defects that cannot be mended are cleaned, and the bone is drilled in order to stimulate new growth of cartilage. This is called the "micro-fracture technique". This has an advantage of not requiring a graft and has overall good clinical outcomes. The only down side is that the cartilage that forms after microfracture is fibrocartilage (scar cartilage), not true Hyaline cartilage (joint surface cartilage), and it has different properties than the natural joint surface.

When the option is to harvest bone that's covered with undamaged articular cartilage from a non-weight bearing part of the body. Plugs of bone (like hair plugs) are moved to a new location. Lastly, cartilage cells can be harvested and grown in a laboratory to make healthy cartilage, which is then implanted. If the cartilage defect is very large and is severely damaged, it may need to be replaced. Allograft cartilage that is fresh frozen may be needed. There are also other procedures being developed, for example, a "biological knee replacement". In these procedures, the body's articular cartilage and bone is harvested, mashed into a paste and put onto the part of body without cartilage to grow a new cartilage piece. This last option is very experimental and not generally available now. These last few options are for larger defects and the first option (bone plug grafting), along with microfracture, are considered most frequently.

It must be noted that even if it's treated, joint damage from an OCD lesion can lead to future joint problems (e.g., degenerative arthritis or osteoarthritis). However, with corrective surgery and once the bone heals completely, most people can return to normal activities.

Osteonecrosis

Osteonecrosis (also known as Avascular Necrosis, or AVN) means "bone death". It is very similar to Osteochondritis Dissecans except that the cause is most often vascular in nature, instead of repetitive trauma; it involves the bone more than the cartilage surface at first. Like in OCD, it is most commonly located in the medial femoral condyle. It could occur on the outside of the leg (lateral femoral condyle) or near the upper end of the tibia as well. Osteonecrosis starts when a piece of bone loses its blood supply and starts to die. In time and untreated, loss of joint space or support for the cartilage occurs as the body reabsorbs the dead bone. It could end up involving the cartilage, as in a case of OCD. If the area collapses, it will resemble severe arthritis and be treated similarly. Women are three times more likely to get osteonecrosis, or AVN, than men are. Women over sixty are especially susceptible. AVN is most often the cause of OCD type lesions in adults and is extremely rare in children.

Doctors are at times unsure of how osteonecrosis develops in a given patient, although it's associated with obesity, sickle cell anemia, lupus, kidney transplants and steroid therapy. It could be from coagulation problems, a genetic disorder called Gaucher's Disease or an injury. Many people get it for an unexplainable reason. In those cases it is called SPONK (Spontaneous OsteoNecrosis of the Knee).

When you have osteonecrosis, you may experience pain and swelling. It will get worse after activity. When the bone segment involved is dying, the joint surface over it may start to collapse, causing a lot of pain. In early stages, the diagnosis can be made by a bone scan or an MRI. In later stages, diagnoses can be easily made by an X-ray, particularly a standing knee film.

A bone scan is a special study that takes advantage of the element Technetium's propensity to find and stick to bone proportional to the amount of blood flowing through it. A radioactive version of Technetium dye (Technetium 99) is injected into the blood stream. The dye is a very weak radioactive chemical so that the radiation exposure is minimized. Then pictures are taken of the bones with a special camera. This camera, like a Geiger counter, picks up very small amounts of radiation. Areas of bone that are undergoing rapid changes, such as a healing fracture, pick up more dye; bone with no blood supply shows no uptake. Because osteonecrosis is 'cold' (meaning there is no blood flow), a bone scan is the best way to see the lesions in the earliest stages.

Left untreated, osteonecrosis can cause severe osteoarthritis and a complete collapse of the bone. Sometimes it can heal on its own, but conservative treatment is necessary as well as modification of activities. Try RICE (rest, ice, compression and elivation) to help with the pain and swelling. Calcitonin is a nasal spray that slows down bone reabsorption and has been thought to be helpful. Complete unloading of the knee is helpful in SPONK, so using crutches, knee imobilizer and/or an unloader brace may be recommended.

Arthroscopic surgery in mild or early disease with debridement of the loose fragments can clean out the knee and make the bone smooth. The doctor may drill the dead bone in order to reduce pressure on the bone surface or stimulate new blood flow. Bone grafts could also help support the knee while the defect heals. In the end stages of the disease, the joint surface has collapsed and can require a knee replacement. In cases where patients may be too young for a knee replacement, an osteotomy, a wedge-shaped cut in the bone that alters the loading on the involved side of the knee by moving the weight from one side of the knee to the other, may help compensate for degeneration near the osteonecrosis lesion.

Knee

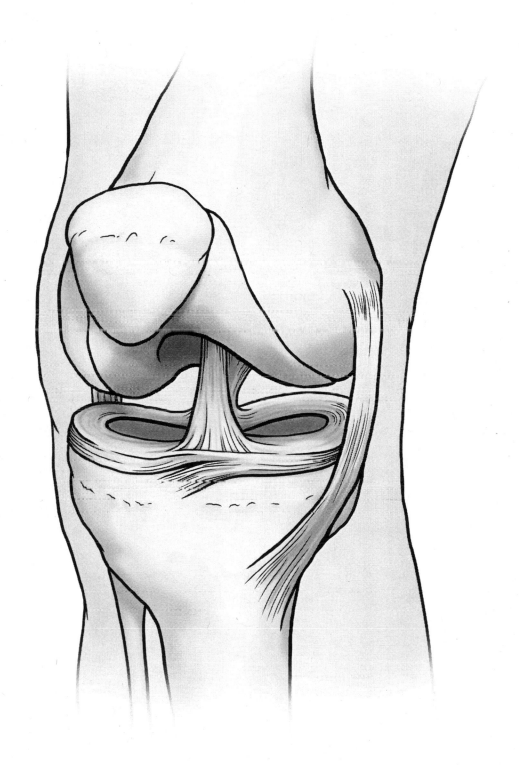

Locking, Buckling and Giving Way

Standing still, I turned to get something off a shelf behind me, and bam, my knee just went." "Every time I get up from a squatting position, my knee won't straighten." "Going down stairs, my knee gives out. I just don't trust it." Frequently, statements like these are the first clue that a patient has an unstable knee. So why does this happen, and what should be done?

The knee is the joint connecting the femur (the thighbone) to the tibia (the shinbone). In the knee joint, the end of the thigh bone is rounded, and the top of the shin is relatively flat. The two bones are very much like a rolling pin sitting on a narrow table. Given even a small push, the rolling pin will fall off. That's why the knee's cartilage and ligaments are so important. They hold the two together and still allow the knee to bend and straighten smoothly. Without the ligaments and the cartilage, we wouldn't be able to run, jump, twist, turn, squat or pivot; and it's when they are injured or not working properly that we have problems.

The examples above are stories of locking (the knee gets stuck in one position and won't move), buckling (the knee is made unstable by a twist or a turn) and giving way (the force of a routine activity causes the knee to stop supporting the body's weight).

Locking can be caused by a piece of torn cartilage (the meniscus) stuck between the bones. Until it's pushed back into place, the knee remains locked and often difficult to straighten. This can be both painful and disabling.

If an examination is positive for signs of injury to the cartilage, a tear may be the reason for the problem. A Magnetic Resonance Image (MRI) or a diagnostic arthroscopy (looking into the knee with a fiber optic telescope) can show the cartilage tear so that the problem can be treated properly.

Giving way can be caused by a cartilage tear or a ligament problem. This is where the physician's examination of the knee is key. Telling the difference can be difficult. This is especially true if the knee is swollen or painful, both common findings in a recent injury. Fortunately, there are specific clinical tests, parts of a good routine knee exam, to help us find the cause. Sometimes, special instruments like the KT-1000 (a very sensitive knee testing device that allows us to measure small movements between the femur and the tibia) can help us decide if one of the major ligaments, like the ACL (anterior cruciate ligament) or PCL (posterior cruciate ligament), is damaged.

Buckling can be caused by cartilage problems, ligament injuries or kneecap problems. The kneecap is part of the quads mechanism. This muscle and tendon unit allows us to kick, jump and squat. It also prevents us from falling when going downhill or down stairs. The body can sense when the kneecap is going to hurt and frequently causes the quads mechanism to release or give (hence the term give-way) to protect itself, and you, from pain.

Once your doctor makes the diagnosis, the treatment for these problems varies. They can include simple exercises, physical therapy, bracing and arthroscopy (fiber optic, outpatient surgery). The early correction of these mechanical problems can lead to a speedy recovery greatly reducing the risk of recurrence, future injury, long term problems and early degenerative arthritis.

"Water on the Knee"

Many times, patients will show up in my office with seemingly unexplainable swelling of the knee. When asked, there was nho history of a fall or a twist. There was no trauma and no history of sports participation. In short, there is no mechanical reason (e.g., a ligament tear, unstable kneecap, torn cartilage, trauma or fracture) for the swelling. In these cases, we must look elsewhere for the cause of the problem.

To better understand the other sources of knee swelling, we must understand what makes the knee, or any joint, move so smoothly. The knee is well lubricated by a constant production of small amounts of fluid made by the lining of the joint. The lining is called the synovium, and the fluid is therefore called synovial fluid. The fluid is what enables the smooth motion of the joint. Conditions of the knee that irritate the lining cause it to make more fluid. If there is enough extra liquid in the knee, it will swell, hence the expression "water on the knee". So what causes the lining to be irritated and make fluid, and, more importantly, what can we do about it?

Why does my knee make fluid and swell?

Most commonly, the knee makes fluid after an injury in an attempt to "solve" the "problem". Like in any injury, if the knee is arthritic for any reason, it tries to fix the lack of smooth motion by making fluid. Therefore, the first cause of "water on the knee" can be just simple wear and tear (in the absence of trauma, a torn ligament or a cartilage tear). Removal of some of the fluid can help decrease the symptoms of water on the knee and yield important clues to its cause. In simple wear and tear, the fluid is clear yellow and a little like a very light syrup in consistency. In other conditions, it may appear cloudy, bloody or opaque. It may contain a high number of white blood cells, altered sugar and protein content, flecks of cartilage, crystals or bacteria. Sometimes, when the diagnosis is not clear, sending a sample of the fluid to the lab for testing is very helpful in finding the cause of the problem.

The lining itself can also be a direct cause of swelling. For example, if the knee accumulates crystals as it does in Gout, the crystals will cause the lining to become inflamed and

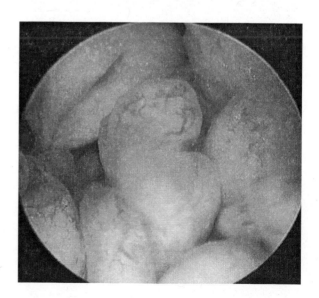

Severe swelling of the lining from psoriatic arthritis of the knee

make fluid, like sand in your gearbox. In Gout, there is a significant inflammatory response, and the synovium and white blood cells try in vain to digest the crystals. Unfortunately, they cannot, and in the process the white cells release destructive enzymes that slowly destroy the joint itself. In Rheumatoid arthritis, the swelling is caused by the body's own immune reaction to cartilage, and the knee makes fluid. The lining in rheumatoid arthritis, as in other inflammatory arthritic conditions, can grow and worsen the situation. There are a whole host of diseases like this, and they can all be painful, cause swelling and ultimately the destruction of any joint that they involve. These include Psoriatic arthritis, that can be associated with even the smallest patches of psoriasis, Rider's syndrome, Anky-losing Spondylitis and Psuedogout (calcium pyrophosphate crystals instead of calcium urate) to name a few.

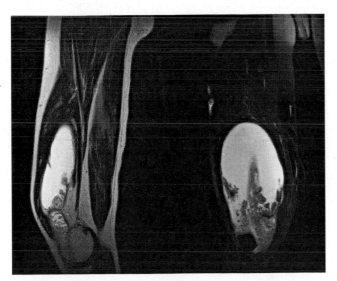

Side and front MRI views of the same psoriatic knee. Note, the fuild shown as white on MRI and the dark noduals in the joint seen on both views

Infection is another cause of swelling. After a viral infection, there are syndromes associated with transient swelling of a joint, a viral synovitis. These are often self-limited. There are arthritic conditions associated with infection, like Lyme disease. This is the second phase of a three part infection that starts with a local rash near the bite of a very small deer tick. In time, the spirochete infection can move to a joint and is associated with a relatively painless swelling. It may resolve without treatment, but that doesn't mean the infection is cured. Un-treated, it will return in its third phase, which in-volves the central nervous system. In this form, it can include seizures, heart rhythm abnormali-ties, coma and even death. Needless-to-say, it is important to recognize potential Lyme cases and treat them in one of the first two stages. Bacteri-al infections can occur after puncturing the joint with a small object. This can be a piece of dirt after a fall on the ground, a nail, a needle or a splinter. These can be urgent problems, and are associated with fevers, painful motion, signifi-cant swelling and redness.

Bleeding into the joint can also be another cause of knee swelling. In special situations, bleeding can happen with a minor trauma. People prone to this are on blood thin-ners, have bleeding disorders like hemophilia or sickle cell disease, have low platelet counts or poor platelet function. Sometimes chronic use of non-steroidal anti- inflammatory drugs (NSAIDs) in high doses or cancer chemo-therapy can inhibit platelet function as well. Recurrent bleeding into a joint without one of these problems can be the only sign of another disease of the knee lining, **PVNS (P**igmented

Cortisone Injections

If a steroid injection has been recommended for your condition, there are several things you should know about the use of these injections. There are many myths about their use, and the rumors are so common that many of my patients have an unwarranted fear of them. In the majority of cases, the proper use of these injects is both safe and effective for many conditions.

For the most part, a steroid injection is a safe, reliable method for resolving the inflammation, reducing the swelling and decreasing the pain of an affected area. Cortisone is a powerful anti-inflammatory drug similar to natural substances produced by your own body. When injected into the affected area, the irritation and inflammation can be reduced dramatically. This can promote both short and, more importantly, long term healing. A misconception is that the injection is simply for temporary pain relief. Our goal is actually long term relief and, combined with the other recommendations, to hopefully cure the condition.

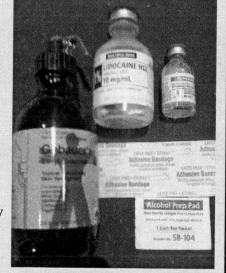

A steroid preparation, as used in our office, is mixed with a short, and sometimes long-acting, local Novocain-type anesthetic (Marcaine). We also add bicarbonate to reduce the acidity of the injection. This takes the sting out and helps the Lidocaine work faster. The injection, therefore, may bring immediate relief to your symptoms and last at least 6-8 hours after it is administered. The cortisone itself takes 7-10 days to achieve its full effect. Therefore, it may take time before your symptoms start to respond. There also may be a few days when your symptoms may worsen before they improve. For that reason, we recommend ice and an NSAID along with the injection and for several days after it.

There is a great difference between injecting Cortisone and taking Cortisone by mouth as a pill. Local injections of Cortisone, in general, have limited side effects on the body, mostly staying locally in the area injected. Still, diabetics may see a transient rise in their blood sugar; therefore, if you have diabetes, monitor your blood sugars and if there is a change, please let us or your diabetes doctor know. Less commonly, some patients can see an increase in appetite, heart rate, flushing of the face or an increase in energy level.

Side effects are rare. One third of patients get discomfort the night of the injection that usually will subside by the next day. Some people can get thinning of the skin or a change of pigmentation; these side effects occur in less than 5% of our patients. You should limit the number of cortisone injections to a given area of the body. In general, I do not recommend more than three injections to the same area in one year. If a third injection is being contemplated, your doctor should be considering some other treatment and making further investigations (like more X-rays or an MRI). Infections are extremely rare after an injection, but even though the area will be prepared sterilely, they can still occur. If there is any concern for an infection (spreading redness, heat or fever), you should contact your doctor's office.

The Cortisone injection is but one component of your overall treatment, which may include non-steroidal anti-inflammatory drugs (NSAIDS) and physical therapy. After your injection, you may resume normal light activities, but you should avoid heavy activity for at least 7-10 days after the injection.

Ice can help diminish discomfort from the injection or the underlying inflammation. Use it 15-20 minutes on, 15-20 minutes off and then on again with some type of thin cloth between the ice and your skin. Caution: Ice left on too long may cause frostbite.

Again, if there is any increased redness, swelling or pain at the injection site, or if you have a fever or chills or other concerns, please call your doctor's office.

ViloNodular Synovitis, with MRI findings simialar to the noduals seen in the example of psoriactic arthritis already shown, except for th additonal finding of hemosidcrin deposits, iron containing, deposits in the noduals).

PVNS is a thickening of the lining of the joint. It can occur in any joint but most commonly appears in the knee. It is seen equally in men and women and most frequently between young adulthood and age forty. It is most often seen in one of three forms. The first is a single nodule of abnormal lining that has hemosiderin (the remains of hemoglobin from blood cells). The second involves a larger area of the lining. The third is considered a locally malignant form that can extend past the joint and into the local tissue. It is very rare (1/2,000,000 of the population) but can be destructive if not treated early and aggressively.

How do we treat non-traumatic knee swelling?

Many of the **arthritic conditions** are treated medically. In mild cases, the inflammations are often controlled with **non-steroidal anti-inflammatory drugs** (NSAIDs) like Motrin, Advil or Aleve. Swelling from mild to moderate degenerative arthritis due to wear and tear of the knee can also be treated with NSAIDs, glucosamine/chondroitin supplements, injections of cortisone and synthetic lubricant injection (Hyaluronan, the same substance that normally lubricates the joint, in a concentrated injection). In cases where weight is an issue, a weight loss program is very important. Most of these patients cannot control their weight on their own. When there is a mechanical deformity of the knee (bowed legs, a valgus knee, or knocked kneed, a varus knee), orthotics, braces and, in some selected patients, corrective surgery is helpful.

In more severe cases, when there is a synovial disease or a **systemic inflammatory condition**, like active Rheumatoid Arthritis, Psoriatic Arthritis or Ankylosing Spondylitis, oral and injected cortisone, immune system modulating drugs (like Methotrexate, Embrel, Anakinra, Remicade and Humira) and/or **Disease Modifying Anti-Rheumatic Drugs** (DMARDs, like Azulfidine, Plaquenil and Arava), along with a rheumatologist's expertise to manage these treatments, are needed.

Acute **Gout** attacks are treated with anti-inflammatories, aspiration and/or injection of cortisone. Long term Gout is treated primarily by lowering uric acid levels in the blood. This is usually done with a medication called Allopurinol and a diet restriction of red wine and protein. Infections that are treated with antibiotics and acute septic arthritis may require emergent drainage as well as intravenous antibiotics.

Blood clotting disorders, like Hemophilia or platelet abnormalities, are most often treated with clotting factors or platelet replacement by transfusion. PVNS is treated by arthroscopic removal of the diseased synovium. The malignant form requires a total synovectomy and radiation therapy to prevent recurrence.

When is surgery helpful?

Arthroscopic removal of the lining (a synovectomy) in some inflammatory conditions has been helpful. When it is done, a biopsy of the lining may also help make the diagnosis, like in PVNS. This procedure is often enough to cure this type of PVNS. Removing all the lining tissue (a total synovectomy) can be required in the treatment of chronic Lyme arthritis. A total synovectomy can also reduce recurrent swelling and pain when the synovitis of conditions such as rheumatoid or psoriatic arthritis is resistant to medical therapy. **Knee replacement** can be considered in the final stages of the arthritic condition, when the cartilage has worn away and quality of life is decreased because of loss of knee function. With modern knee replacement surgery, the surfaces are replaced with metal or ceramic prostheses and often a plastic liner with very high success rates. Like after hip replacement surgery, the patient's satisfaction after knee replacement is amongst the highest of all surgical procedures.

Pre-patellar Bursitis ("Nursemaid's Knee")

There are many bursas in the body. They serve as sliding surfaces between body parts to help them move smoothly. They are more commonly located between a tendon and a nearby bone. There are still some between the skin and bone, near joints that have a good range of motion. When a bursa gets inflamed, it rubs against the bone and swells. There is a bursa between the skin over the kneecap and the kneecap itself. When this particular bursa swells, it looks like there is something wrong inside the knee. Often there is nothing wrong inside the knee joint. This is truly what is meant by "water on the knee." The water is in a sack in front of the patella and not in the knee at all. Most of the time, this bursa swells from continued kneeling and repeated mild trauma to the front of the knee at work, or repetitive motion involving the front of the knee. It was most commonly known as nursemaid's knee. This is because the nursemaid spent a lot of time kneeling (washing the floor), and the swelling was characteristic of constant trauma to the bursa. It is also seen in carpenters, plumbers, carpet layers and other laborers.

Treatment

Nursemaid's knee is treated first with rest, ice and NSAIDs. If those treatments fail, your doctor may try an aspiration of the knee and an injection of cortisone. In chronic cases, removal of the bursa may be necessary. I have had cases where a bursa had become infected after being puncture by a foreign object (like a tack or needle) left on the floor. In one rare case, a few grains of sand were imbedded in the bursa after a fall in the dirt. Years later, the painful swelling refused to clear. This was due to an infection that was so mild, but chronic, because the bacteria that caused it only lived on the grains of sand. It just hung onto the sand and could not spread elsewhere. This was completely cured only after the foreign bodies (grains of sand) were removed.

In patients with occupations that require kneeling, like carpet installers and tile masons, kneepads are necessary to avoid these issues and should be worn at all times. Even those who wear the pads may still have bursitis that is very hard to treat because they continue to traumatize the area daily.

ACL Tears

The Anterior Cruciate Ligament, or ACL, is located in the center of the knee joint. It is one of the four major knee ligaments that control the hinge-like movement of the knee. This ligament prevents the tibia (shinbone) from going too far forward in relation to the femur (thighbone) during activity. Some consider it the most important ligament in the knee because it helps maintain the stability of the joint during pivoting and cutting activities (fast changes in direction) in sports. In the knee, the ACL sits in front of the Posterior Cruciate Ligament (PCL), and they physically form a cross. This is

Conceptual drawing of the cruciate ligaments.

why they are called "cruciate" ligaments, which comes from the Latin word meaning "cross-shaped". A tear of the ACL is one of the most common injuries in the knee, and it is especially a problem for female athletes (1.4 million women have torn their ACL in the last ten years, see "Female Athletes").

The ligaments in the knee give it a certain set range of motion. A tear, or sprain, occurs when the knee is forced past that range. Most ACL tears come from twisting injuries.

They could also occur from the tibia being pushed too far forward or the leg being suddenly over-straightened. A tear is most likely to occur during sports such as basketball, football, skiing and soccer. In those cases, the leg may be twisted while the foot remains planted, hyperextended in a direct hit or quickly forced forward by the back of a ski hitting a mogul top. An injury can also happen while running. The runner could turn too suddenly, stop short or land incorrectly from a jump. After a tear, the tibia can shift forward more freely and the knee will buckle more easily.

How do you diagnose an ACL tear?

At the injury, you will feel a pop or like something has snapped. It is usually not that painful. The knee will often swell within two to four hours. The ligament is vascular (meaning it has many blood vessels), so the small vessels in the torn ligament will slowly bleed into the joint causing a hemarthrosis or blood filled swelling of the joint. Once the knee is swollen, you may experience pain from the joint being tight inside. With the ligament torn, you may also experience an inability to walk or giving out and buckling.

Your doctor can often make the diagnosis by taking the history and performing a physical exam, which may include a Lachman test and a pivot shift test. In a Lachman test, the anterior movement of the knee is tested by pulling forward on the tibia while holding the

Female Athletes

Female athletes have a much higher chance of tearing their ACLs than men have. The NCAA found that female athletes were six times more likely to tear their ACL than male athletes each hour of play. Some call ACL tears in female athletes an epidemic. Why is this so?

Women are anatomically different than men. First of all, they mature a lot faster. The quality of bone and its stiffness changes when a person stops growing, making the bone less forgiving and the ligament more susceptible to tear. Furthermore, women have wider pelvises and legs that, on average, tend to "bow outward" (valgus alignment) as opposed to "bowing inward" (varus alignment), like men's legs. The roof of the center of the knee, called the femoral notch, is narrower in women than in men, placing more strain on the ACL with knee extension. Female hormones also seem to cause increased joint laxity of the "secondary restraints" of the knee. These are ligaments that help guide knee motion along with the ACL. When they are less taut, they put more stress on the ACL. The early age of maturity, leg alignment, tighter space around the ACL and hormone-related looseness of the knee are all considered important reasons for an increased injury rate in women athletes over their male counterparts.

Women's and men's muscles react in differing ways during sports. Video tapes of high school athletes during competition have taught us that men and women have their own ways of jumping and landing. Women tend to bend their knees less, landing with their knees straighter, and they use their quadriceps muscles to a greater extent than their hamstrings. This causes them to be more flat-footed when they land. This lack of muscle balance causes increased strain on the ACL, particularly at the end of a quick move. Adding the muscle pattern difference to the anatomic factors explains why many of the injuries occur as non-contact events.

Since the anatomic factors are unchangeable, a woman can decrease her rate of injury by working on improving her muscle balance and landing style. From the video tape studies, we know that practicing two-legged landing has the potential to reduce the risk for female athletes. Keeping their knees more bent while jumping and landing on the balls of their feet while using their hamstrings (back of thigh) and glutei (buttock) muscles will help protect their knees. In fact, studies have shown that preseason and in-season balance and jumping programs provide some protection, although not all authors agree on how much. Rehab can help, and the leg muscles will compensate for some of the loss, but they cannot replace the mechanical protection that the normal ACL gives in high level sports. The result is often recurrent injury that leads to tearing of the meniscus or other cartilage. Damage to the cartilage can be the start of arthritis and is associated with the knee locking (getting stuck in one place). With repetitive injury, the knee can become completely locked, and the player will not be able to extend the knee or even walk on it. To avoid these long term problems, most injured athletes opt for surgery.

femur in place. In a pivot shift test, the doctor will test the knee's rotation as it bends. Both of these tests are meant to simulate what happens when the knee gives out. An X-ray may show a small fracture at the lateral edge of the tibia that could be present with an ACL tear and confirm the diagnosis. Furthermore, an MRI can confirm the tear in the ligament and reveal any other damage to the cartilage or bone. A KT1000 test, an instrumented Lachman test, can also be used to confirm the diagnosis and help determine the best treatment choice.

What is a KT1000 test and how does it work?

A KT1000 is a little machine that is strapped onto the tibia. One pad touches the tibia and the other touches the kneecap. These pads are used to measure the range of motion in the leg, testing the ACL by pulling up the tibia. A typical reading for a normal leg in this test would be around 8 mm of anterior displacement. With the knee tester in place, the patient can be asked to contract his or her quads, which helps determine if the PCL is torn as well. The normal leg will be tested first, and then the injured side is tested to see the difference between the two. This is called the "side-to-side" difference. If the ACL is torn, the displacement measured for the injured knee will be larger than the reading for the normal side.

Normal knees have a side-to-side difference (injured minus normal) of less than 3 mm, so if the difference is greater than 3 mm, the ACL is diagnosed as unstable or torn. Sometimes the ACL can be torn, but the remaining uninjured ligaments still keep the knee stable. In those cases, patients will still get a side-to-side difference of less than 3 mm. These patients may do well with non-operative treatment.

Dale Daniel (one of the inventors of the KT1000) and others have shown that the level of instability in the knee can predict the risk of future injury with respect to activity level. For example, a patient with low demand and a non-athletic lifestyle may have very little difficulty with a side-to-side difference of 5 mm. However, a high demand athlete may have difficulty with this same side-to-side difference. Furthermore, patients with more than a 7 mm difference will have problems with activities of daily living. This instability has been shown to increase the risk of cartilage tear in the knee, recurrent swelling and, as a result of the cartilage injury, subsequent arthritis.

The KT1000 helps us decide whether people need surgical repair or not. It has an advantage over an MRI since it's a physiological test and measures the ACL's actual function. Sometimes an MRI shows swelling in the ACL or a cyst in the ligament, but the ligament fibers are actually intact. Conversely, if a patient has a locked knee, where the meniscus is trapped in the middle, the KT1000 test may appear normal even if the ACL is not working. The combination of using an MRI and a KT1000 exam helps to better determine the best treatment for all patients when the ACL is injured.

Nonsurgical treatment of the ACL

If the knee is stable enough, nonsurgical treatment is suggested. The patient should try RICE (rest, ice, compression and elevation)

Training the Brain to Prevent ACL Tears

New research has sh`own that the best method for ACL tear prevention may actually be to train your brain. When athletes are tired, their reactions to unanticipated commands become slower and more dangerous, making tears more likely to occur. In the latest research, this is also true when only one leg is fatigued and the other is left alone. The tired brain (from fatiguing activity) reacts slower, and even the limb that has been spared from the fatiguing exercises in the test protocol exhibits poor muscle reactions. The study proposed that this is because the brain cannot react well even though the muscle is okay. Therefore, in this case, both limbs are at risk for an injury even though only the muscles on one side have been made "tired".

The researchers concluded that training the brain and a person's reflexes may be the best way to counter this issue. By exposing the athlete to many controlled movements and training his or her responses for good balance and muscle control, it may be possible to improve reaction time in sports. By sharpening an athlete's anticipatory skills, one may see a reduction of injuries. Some studies have shown this to be true. When mentally prepared, the athlete can avoid dangerous actions, react better when they occur and help avoid those that lead to injury.

and NSAIDs. If he or she continues loading the knee and the knee gives way, the meniscus can become damaged. Therefore, crutches should be used for a few days after the injury. Ice and compression will help relieve the swelling, giving back the full range of motion. Wearing an ACL brace can help prevent re-injury. In some cases, if the knee is stable, a brace may not add any stability, but it may add another way for the body to be able to feel the knee's position (the

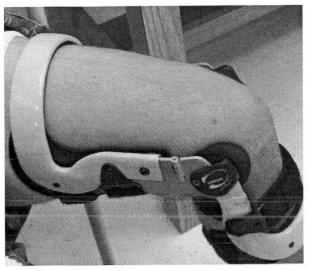

ACL Unloader Brace

sense called proprioception) and therefore be protective. When your body knows where your knee and foot are, it will be able to prevent you from falling by using the correct muscles during activity. Physical therapy is important to help strengthen the muscles around the knee. For a return to sports, you might consider conditioning and bracing (see "Female Athletes").

Surgical repair of the ACL

ACL surgery is suggested if people don't respond to nonsurgical treatment, or it is initially recommended based on understanding the relative instability of the knee (side to side difference), the level of demand the patient places on the knee, other injuries present (e.g., associated meniscus tears) and the age and activity level of the patient. It must be noted that if a repairable meniscus tear is present, surgery is recommended since the outcome of a meniscus repair is improved when the ACL is reconstructed at the same time. The ACL tear repair surgery is done as an arthroscopic outpatient procedure. A tendon graft can be taken from the patient's patellar tendon, quads tendon or hamstrings (called an autograft) or it can be taken from a

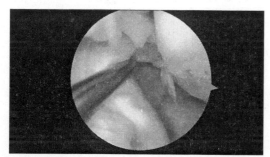
An arthroscopic view of torn ACL fibers.

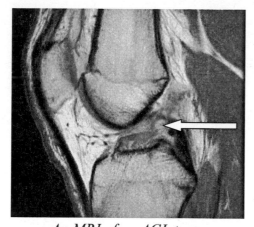
An MRI of an ACL tear.

donor's patellar tendon, hamstrings or Achilles tendon (called an allograft). Bone tunnels will be drilled into the tibia and femur. The graft is then guided through the tunnels and can be anchored with titanium or bio-absorbable screws, buttons, pins and other devices. After the graft

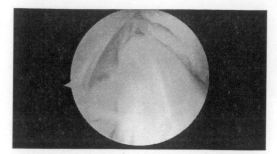
An arthroscopic view of an ACL after repair.

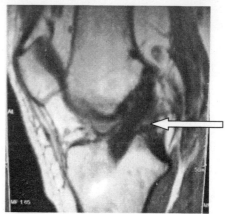
An MRI of an ACL 5 years after repair.

is in place, the small incisions are closed.

There are a few different options for surgery. The most common autograft is from the patellar tendon, but some doctors choose to use a hamstring tendon (a semitendinosus or gracilis) autograft. Because of past complications with synthetic grafts, most surgeons do not use synthetic substances for reconstruction anymore, although there are some new experimental grafts being tested (that are not currently approved). Using allografts creates the advantage of not having to use another part of the patient's knee to repair the damaged ligament. There have been a number of reports of the effects of losing the donor site tissue and complications related to the healing process. For that reason, allografts have gained in popularity.

There are many factors that help decide between using an autograft or an allograft. Autograft donor sites can be painful after surgery and the donor site may take a long time to heal. Also, the patient is at risk of getting tendon weakness at the graft donor site. Allografts are less traumatic in that no normal structures are taken from the injured knee, and they have a high success rate. However, there is a small chance that the body will have a reaction to the allograft (rejection does not really occur because there are very limited cellular elements), and the graft needs to be tested for communicable diseases before the tissue bank can release it for use. Some doctors prefer one type over another and will therefore favor that type of surgery. If they are more skilled in one area, it is wise to pick that option. Remember to trust your surgeon's judgment.

It has also been noted that female athletes are at an increased risk for valgus collapse. When the athletes jump and land, the knee falls into a knocked-knee position, and this increases stress on the ACL. Along with the other factors mentioned in this chapter, the knee is at risk for injury. There is some data that shows that training can decrease valgus collapse. More recently, special shorts have been designed to help guide the legs and decrease valgus collapse when used in conjunction with a special training program (see www.Aspaeris.com).

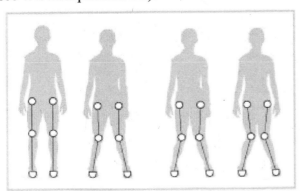

Reprinted with permission from Aspaeris.

The diagram on this page demonstrates common postures observed while female athletes are in motion. From left to right, the diagram depicts increasing degrees of "knocked-knee," scientifically known as valgus collapse. It is this dynamic valgus collapse that may hinder performance and cause injury. Performing these simple exercises below can test for dynamic valgus collapse:

Performing a squat

- Have someone stand directly in front of you (your "spotter")
- Stand with your legs hip-distance apart
- Squat deeply like you are sitting in a chair, repeat 10 times in a row
- If the spotter sees your knees shift inward, you have some degree of valgus collapse

Performing a jump shot

- Have someone stand directly in front of you (your "spotter")
- In front of a basketball hoop, perform a basic jump shot
- If the spotter sees your knees shift inward, you have some degree of valgus collapse

If the athletes fail these tests, they should enroll in a balance improvement and ACL prevention exercise program. Although long-term outcome studies are not available yet, protective garments like Aspaeris control shorts for female athletes may also decrease valgus collapse and act as a prevention aid in these cases.

In a final note, the American Orthopaedic Society for Sports Medicine (AOSSM) has recently conducted a review study of the risks and benefits in repairing the ACL of a child under the age of fourteen. Usually, a doctor would wait a while before operating, for fear of creating a growth problem. This study shows that it is actually more prudent to perform the surgery sooner rather than later. The patients who wait for surgery are more likely to have additional injuries and knee problems because the knee is left unstable. These include: irreparable meniscus tears, chondral injuries and patellotrochlear injuries. From the current studies, it appears that delaying an ACL repair in young patients may create more risks than benefits.

Injury to Other Knee Ligaments (MCL, LCL, PCL)

The ACL (Anterior Cruciate Ligament) gets so much press that other ligament injuries seem so much more mysterious and far less common. From the NY Jets star quarterback, Joe Namath (one of the first athletes to have a major knee reconstruction in the 1960's), to golf's Tiger Woods (in 2008), after nearly 50 years, the ACL still gets all the headlines. The truth is, the knee has four major ligaments, two menisci and a curved shape, and they all play an important role in its function. The four ligaments, ACL, PCL, LCL and MCL, all help to stabilize the two bones of the knee while allowing for full knee motion. It is a marvel that it works as well as it does. Knee ligament injuries vary widely. As we have already discussed, the ACL can be injured with hyper-extension, pivoting and with quick cutting motions.

MCL

Of all the knee ligaments, the MCL (Medial Collateral Ligament) is most injured and least reported. This may be due to the fact that it is the strongest of all the ligaments (a little known fact), and it has two layers so that partial injuries or sprains heal quickly and leave little functional deficit while healing. The ligament is located in the inner side of the knee, outside the synovial lining of the joint. Still, it can be sprained, partly torn or completely torn (grades 1, 2 and 3). The exam will show little or no laxity in grades 1 and 2 to valgus stress.

Pain to stress testing and palpation over the femoral or the tibial insertion sites help to make the diagnosis. Grade 3 sprains are far less fun, as they are complete disruptions which leave knee unstable. They require bracing for six to eight weeks and even more time to regain full strength. When torn, the proximal MCL injuries heal better than the distal MCL. Occasionally, a severe grade three sprain fails to heal. When a grade three tear does not heal and instability persists, reconstruction or reefing is needed. Sometimes reconstruction with a tendon autograft or tendon allograft is used to reconstruct the loose structures. Children have an added concern, since injury to the growth plate can masquerade as an MCL injury. If an MCL is suspected in a growing child, X-rays including stress views, or possibly an MRI, are mandatory to be certain that the growth plate is not injured. Growth plate injuries in children require complete cast treatment and/or other strict immobilization.

LCL

The MCL tears far more often than the LCL (Lateral Collateral Ligament) tears, even though the LCL is the weakest of all four ligaments. The strength of the knee ligaments in order is MCL > PCL > ACL > LCL. Imagine how easy it is to be hit on the outside of the knee during sports. A blow to the outside of the knee is what stresses the MCL. Stressing the LCL requires the opposite stress, a blow to the inside of the knee. This injury is far less likely to occur in all sports than a blow to the outside of the knee. Therefore, even though the LCL is far weaker, it is much harder to injure. When the LCL is torn, it is often associated with tears of the lining of the knee joint on the same side (the posterior lateral corner of the knee). This causes rotational instability and, on examination, the foot will turn out more on that side when compared to the uninjured knee. This extra rotation is called a positive dial sign. When repairing the LCL, it is recommended that the repair be augmented with a graft to help resolve the rotational instability.

PCL

The PCL (Posterior Cruciate Ligament), like the ACL, is inside the knee joint. It is stronger than the ACL and is injured less frequently. The injury is often from a direct blow to the shinbone, a fall onto the leg with the foot pointed down or a dash board injury when the shinbone hits before the kneecap. These patients have swollen knees that improve with time. With an isolated PCL tear (all the other ligaments are intact), people do not complain of knee instability or giving way like ACL patients. They usually have a slow onset of increasing kneecap pain or medial knee pain. This is primarily because of increased stresses seen by the kneecap and medial knee as a direct result of the loss of the PCL's main function (holding the tibia forward with knee motion). The slipping back adds load to the only structure that can pull the tibia forward, the thigh (quadriceps muscle) and the kneecap. The shifting back and forth wears the medial knee too. Most experts recommend non-operative treatment for isolated PCL tears with limited posterior sag or translation (less than 10 mm) and repair for translations greater than 10 mm. With translations greater than 10 mm, it is important to check for other injuries. They are often the major reasons for the increased instability.

Torn "Cartilage" in the Knee (Meniscus Tears)

The menisci are gasket-like cartilages that sit on top of the tibia (shinbone). Each knee joint has two menisci. Together, they fit snugly around the rim of the knee. They act as a cushion for the femur (thighbone) on the tibia. They work to absorb shock and, with the articular cartilage, they prevent the leg bones from grinding together. It was once thought that the menisci were left over from development and of no use. It is now well known that they can carry up to 50% of the force across the knee, and if they are completely removed, arthritis will develop. The lateral (on the outer side of the knee) meniscus is semicircular. The medial (on the inner side) meniscus is C-shaped and is more likely to be injured. Once a meniscus is injured, the patient may experience locking, buckling and giving way.

Tearing of the meniscus can occur in any age group. It usually results from forceful twisting of the knee with the knee loaded (weight put on the knee) or planted on the ground. It can also occur with squatting or hyper-flexion of the knee, after jumping and landing with the knee flexed or from twisting as the foot hits an obstacle, like a hole, rock, uneven ground or object. In addition, sometimes a tear can occur from a very minor activity when there is a previous injury or a predisposing factor present. Tears can also come with trauma that causes damage to the other ligaments. An injury to both a ligament and the meniscus

A degenerative tear.

makes the knee very unreliable, causes swelling and usually prompts a visit to an Orthopaedic Surgeon.

A younger person's meniscus is tough and rubbery, and it is most likely to be torn in a sports injury. The menicus may soften with time and, for more mature people, the tear can occur from squatting, twisting or falling. In older patients, the meniscus can be even weaker and thus easier to tear. Degenerative tears of the meniscus can be seen as a part of osteoarthritis of the knee, gout or other arthritic conditions. In those cases, the damage is more gradual and the symptoms will appear over time. Repetitive compressive or rotational stress on the knee can cause the bones to strain the cartilage, causing further wear and tear. Finally, pieces of the meniscus can come off and go into the joint; these are called loose bodies. They can cause pain and swelling as well.

How will I know if my meniscus is torn?

Usually there is no one associated injury that leads to a meniscal tear. As states a twist and pop are common followed by swelling and knee pain. The knee may lock and be hard to straighten. Pain may be felt along the joint line where the meniscus is located, but sometimes the symptoms are vague and occasionally involve the whole knee. Pain and swelling may not occur until the day

The meniscus begins as a disc shaped structure. As the knee develops, it changes to the more common "C", or semi-lunar, shape. On occasion, the lateral meniscus does not change and will retain its "discoid" shape. People with this condition may never become aware of it, but sometimes it can be unstable and tear. When the damaged discoid meniscus causes locking, swelling or pain, surgery may be required to treat the problem. If the patient has surgery to repair this, the surgeon may remove the torn part and reshape the meniscus into the more natural "C" shape.

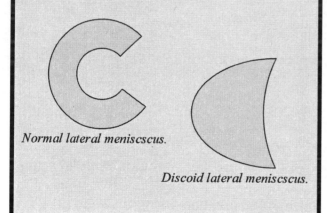

Normal lateral meniscscus.

Discoid lateral meniscscus.

motion) or giving way (locking or catching while walking, causing the knee to stop supporting the body's weight). Less commonly, the knee becomes locked in a bent position and cannot be straightened at the time of inital injury. This happens when the torn part of the meniscus is large enough to get stuck between the bones in the hinge mechanism of the knee, causing a true block to knee motion. Walking is dificult and running is almost imposible with a locked knee. A "locked" knee requires surgery to reduce and repair the trapped, torn meniscus.

In general, even in knees that are not locked or only lock intermentantly, if meniscus tears are left untreated, they can cause constant rubbing of the torn section against the tibial or femoral surface. In time, this will cause damage or degeneration of the knee joint. The knee may become swollen, stiff and tight.

To make the diagnosis, your doctor will ask for the history of the injury and examine your knee. He may test your knee in bending, check for swelling (an effusion), check your ligaments, check for joint line tenderness (pain over the tear when touched), perform a McMur-

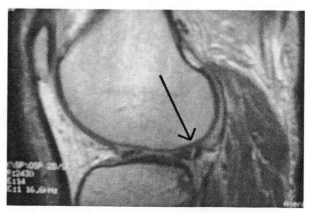

A lateral meniscus tear.

after the injury. You may experience pain when going up and down stairs, squatting or getting up from a low seat. You may feel the sensation of giving way. You may also hear a clicking sound when you move your knee.

After the initial swelling goes away, a patient with a meniscus tear will complain of recurrent swelling without warning, the inability to squat or kneel, locking (the knee gets stuck, requiring some twisting to unlock it and restore ray's test (twisting your knee while bent and seeing if straightening it causes a pop or pain), an Apley's compression test (twisting the knee when loaded to see if there is pain) and/or a flick test (pain when flicking the knee in short, quick rotation).

Treatment options

Once torn, the meniscus will not heal on its own. There are two options for surgery: either removing the torn portion or repairing it. The choice in surgery depends on the location and type of tear. The meniscus is made up of a "red zone" and a "white zone". The red zone has blood flowing into it. The white zone does not. If a tear is "red on red", meaning both sides of it are in the red zone, it is repairable. If it is "white on white", there is no blood supply, and it is not repairable. If it's in the "red on white" zone, it is sometimes repairable. A clot can be made with the patient's blood to aid in the healing process of this type of tear. Many times, the meniscal surgery is performed during another surgery on a ligament in the knee, since meniscal tears are usually associated with ligament injuries. The success rate of a meniscal repair increases when an ACL tear is repaired at the same time.

If the tear is small, stable, in the red on red zone and in a young patient, it can heal on its own. RICE and NSAIDs should help, and crutches are recommended so that you don't put pressure on the meniscus and worsen the injury. Wearing a brace helps as well. A good physical therapy plan is important. If the tear is small, it may heal in four to six weeks. Larger tears

with an unstable knee will not heal. If a patient experiences locking, giving way or recurrent swelling, the tear is unstable.

In the cases in which repairing the cartilage is impossible, removal of the torn fragment is necessary to return the knee to good function. This is called a partial meniscectomy. For example, in older patients, degenerative type tears are usually unrepairable. Any loose pieces will be removed along with the torn portion, and the surface will be left smooth. This is called debridement. It must be noted that even removing only parts of the meniscus will create some increased risk for arthritis. This is because the

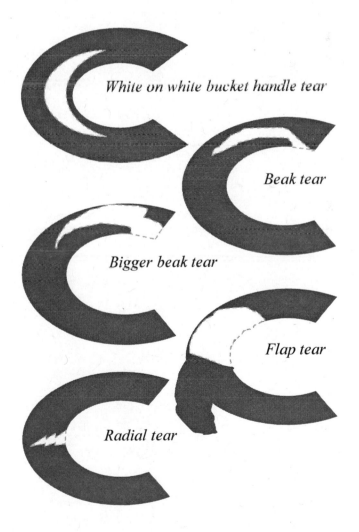

White on white bucket handle tear

Beak tear

Bigger beak tear

Flap tear

Radial tear

bone's cartilage cover is under higher levels of stress without part of the meniscus acting as a cushion. Of course, even if the tear is irreparable, some meniscus is better than no meniscus and our goal is to always keep the parts that are functioning. Below are examples of types of

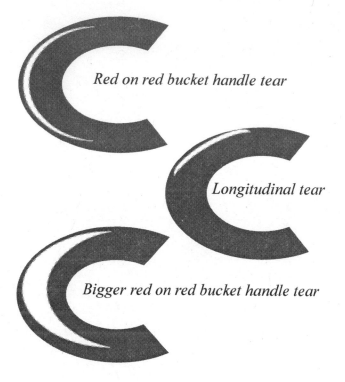

Red on red bucket handle tear

Longitudinal tear

Bigger red on red bucket handle tear

tears that are repairable:

When the meniscus can be repaired, tiny sutures or stitches are used to fix it arthroscopically. This is called a meniscal repair. The ability for repair depends on the location and also the condition of the cartilage. These are examples of types of tears that are repairable:

The healing rate after surgery is inversely proportional to the tear length, the patient's age and the location of the tear. The shorter the tear is, the younger you are and the more blood flow there is, the faster it will heal. In repairable tears, roughening the edges that cause bleeding, stimulating the bone marrow, stabilizing the knee (an ACL reconstruction) and adding a

Meniscal and Baker's Cysts (Popliteal cysts)

Occasionally, a meniscal cyst may accompany a tear. The tear is an irritant to the knee, and the knee will react by increasing its lubricant and making more joint fluid. The fluid can be trapped by the tear and cause a local cyst that communicates with (is connected to) the tear. In practice, medial cysts occur more often than lateral cysts. Lateral meniscal cysts are harder to treat than medial meniscal cysts, and they have a higher risk of recurrence. The fluid can also pump into the back of the knee into the popliteal space. This fluid collection is then called a popliteal, or "baker's", cyst (baker's often had these types of cysts from many hours of standing). In either case, a sac of fluid forms. It can be recognized as a hard lump on the joint line or a sense of fullness felt in the back of the knee. To reduce the symptoms, the cysts can be aspirated or drained. When a baker's cyst becomes large enough, it can rupture, causing pain and swelling down the calf. If it does rupture, it may feel like warm water running down the back of your leg. Large baker's cysts can also disturb the circulation to the foot and cause foot and ankle swelling. A cyst won't usually completely resolve unless the tear is treated by removal or repair. Once the tear is treated, it should go away on its own. If it doesn't, draining it may help. If it still does not go away, the cyst may need to be surgically removed.

fibrin clot or activated platelets all help in the healing process.

It is well known that complete loss of the meniscus leads to arthritis. Therefore, replacing the meniscus when it is lost may be of value in a few select patients. Meniscus replacement with an allograft is a technically difficult procedure with a long rehab period, but it is considered at times. Most recently, meniscus scaffold material has been approved by the FDA, and it has been used in Europe for more than 5 years with some success. This may also be an option for those that have had more traditional treatment, like a partial meniscectomy, and continue to have symptoms. Both of these procedures require careful patient selection, long rehabilitation and significant surgical

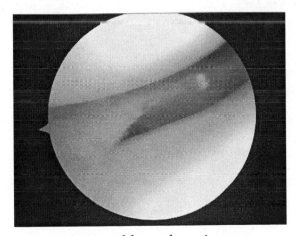

A normal lateral meniscus.

expertise.

In the end, the treatment for meniscus tears can be summed up by a few simple principles: Keep want you can, repair what you can, protect a repair from re-tear while healing and if you do a repair, create the best healing environment in the joint as possible (e.g., add a fibrin clot or do an ACL reconstruction if needed).

Meniscus scaffolds and allograft meniscal transplants

At times there are patients who have had a partial meniscectomy but continue to have medial symptoms. This is despite having a good surgery and being in reasonable knee alignment. These patients may be candidates for restoration of the missing meniscus fragment or allograft transplantation of a new meniscus. For a meniscus allograft, size is very important, and a good make in size is the first priority for surgery. The second surgical goal is secure fixation of the bone ends of the meniscus allograft. In cases with a great fit (the size is right) and great fixation, the meniscus graft heals to the rim (90% of the time, which is better than many meniscus repairs), and the patients seem to improve. More recently, the FDA approved a meniscus cartilage scaffold device. This is a pre-shaped collagen matrix that can be cut to shape and sewn in place. The five year data from Europe (published in JBJB in 2009) shows that in selected cases, it improves function and decreases symptoms.

Other Cartilage Defects

In the knee, there are two types of cartilage: meniscal and articular. In this chapter, we are going to explore issues concerning the less well-discussed, but still very important, articular cartilage. The articular cartilage can be found on the surface of every joint in the body. In the knee, it is the specialized cartilage surface that covers the ends of the femur (thighbone), patella (kneecap) and tibia (shinbone). It is the smooth covering that makes the surface glide freely. It also serves as protection for the bone and as a shock absorber. As smooth as it is, it can still wear or become damaged. It can be worn by the repetitive stress that's put on knee by heavy activity. It can also be damaged during trauma or ligament injuries. Once there is a damaged area, the surface is prone to degeneration since it will not heal on its own. Sometimes the cartilage is softened, and occasionally the bone underneath the cartilage is damaged, removing its natural support. Sometimes the cartilage defects are shallow and sometimes they are deep. All of these conditions can cause knee pain and swelling. This ongoing damage and loss of the smooth surface can lead to arthritis. When these defects are symptomatic and they cause pain, locking or recurrent swelling, surgery is often needed to improve knee function.

Shallow cartilage defects may be different than Osteochondritis Dissecans (see previous section) and are more common in adults. They may be caused by trauma, wear and tear or arthritis. These defects can be treated non-operatively, with anti-inflammatory medications, cartilage food supplements (glucosamine and chondroitin sulfate), cortisone or synthetic lubricant injections (hyaluronic acid). They can also be treated surgically with simple debridement (removing the loose fragments), the microfracture technique, replacement with synthetic scaffold, transfer of cartilage from another part of the knee (autograft), frozen cartilage grafts (allografts) or an ACI (Autologous Chondrocyte Implantation, harvesting your own cells, growing them on a Petri-dish and replanting them).

Nonsurgical treatment

Non-Steroidal Anti-Inflammatory Drugs (NSAIDs) are the mainstay of non-surgical treatment for inflammation anywhere in the body. In a knee with surface defects, it decreases the swelling and reduces the pain associated with wear. In theory, some of the anti-inflammatory can reduce the swelling in the cartilage itself and "toughen" the cartilage. There are a fair number of these drugs and GI (gastro-intestinal) upset is the most common side affect. They also can increase bleeding times (the time it takes to form a clot). In this sense, many of the drugs can be protective against a clot, like aspirin is. At the same time, they (as a class of drugs) can reduce the stomach's natural protection against acid and cause ulcers. Selective NSAIDs that block only part of the inflammatory pathway can have less of these side effects, but many individuals react differently. Therefore, one drug may work well for you, with little side effects, but not be tolerated well by your neighbor. When selected for a patient's specific needs, these drugs can be both safe and effective for the majority of people.

Glucosamine and Chondroitin (G-C) sulfate are food supplements that are touted to have a wide range of benefits, including re-growing cartilage and restoring the joint space. These are somewhat overstated to say the least. The advertisements that show knee X-rays with improvement in joint space after treatment have been proven false. However, it does seem that there are patients that benefit from taking these supplements. A few studies have shown that up to 50% of patients see some reduction in pain, as well as improvement in function. We are reminded that the placebo effect alone helps 30% of all patients. Just the same, I personally believe (this is my own conjecture here, so don't quote me!) that there are patients that lack the building blocks of cartilage in their diet, and a food supplement does replace that deficit. It may be like chicken soup for the knee. Of course, as an idea this is appealing, but it may not be true. In practice, patients looking for alternatives can try G-C, but if it does not help decrease daily pain in one to two months of treatment, it should not be continued. (For more information on G-C treatment, see the side bar on page 49.)

In an inflamed knee, when these oral agents fail and there are no true mechanical symptoms, a cortisone injection may be of great help (see side bar on injections on page 26). In many cases, patients get good relief from pain and can return to activities within a week to ten days. In some cases, the relief is short lived, and in others, it lasts only a few hours. Patients with short-lived relief may benefit from Hyaluronate (lubricant) injections. If the patient has only a few hours of relief, we can be reassured that the knee is the cause of his or her pain. This is because the knee injection usually contains a local anesthetic, and if numbing the knee gets rid if the pain, the knee is the most likely source of the problem. Conversely, if the knee does not improve at all and the local has no benefit, other sources of the pain must be investigated (like a herniated disc or hip arthritis).

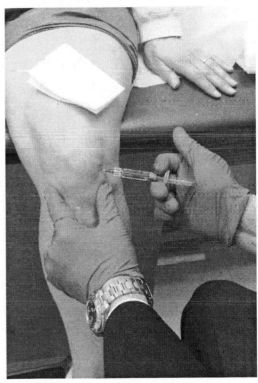

Synthetic lubricant injection.

Every joint makes its own lubrication. The lubricant is mostly hyaluronic acid, a large molecule that forms a thin oily substance that, in combination with the microstructure of the articular surface, creates the extremely low friction surface that can last a lifetime. Many companies make these injections in higher and lower molecular weights and differing purification schemes. The original FDA approvals ranged from three to five injections one week apart, depending on which product you used. Most recently, one company received approval for a single injection form (naturally a larger volume of lubricant). It is being used in some settings. Since it is newly on the market, it is not clear yet how well the single higher volume injection is tolerated in all patients.

Surgical treatment

If there are some mechanical symptoms, an arthroscopic debridement may be of help in some patients with limited cartilage damage and loose fragments in the joint. One must be aware that debridement alone for advanced arthritis has been criticized by some and, in one larger study, was shown to be better than a "sham" operation. Many have taken issue with the one study because it was not clear if there was significant bone loss pre-operatively, and there was no clear discussion of true mechanical symptoms. Nonetheless, when all else fails, the knee is the clear source of the pain and the patient is not symptomatic enough or too young for a knee replacement, a diagnostic arthroscopy may show the cause of the problem, help reduce the symptoms or even correct the problem.

If the symptoms of a cartilage defect are ignored and spontaneous healing does not occur, the cartilage and its base will separate from the diseased bone, and fragments will break loose into the knee joint. If the fragments are loose, your surgeon may scrape the cavity to reach fresh bone. Fragments that cannot be mended

will be removed. Removal of the fragments will leave a defect hole that needs to be repaired. When there are measurable defects in the bone and cartilage is loose, other options should be considered. For smaller lesions, a microfracture technique is used; the surface is cleared of the damaged cartilage, and tiny holes are made in the bone to allow the marrow to "leak" out and cover the boney bed. To accomplish this, the cartilage surface is drilled, or "cracked", with a microscopic drill or awl to help blood and marrow get to the surface, just like aerating the soil when you want more grass to grow. The new cartilage will cover the surface with fresh tissue. The surface defect will fill in with fibro-cartilage in time. Unfortunately, although it helps, it is not the same as hyaline (articular) cartilage.

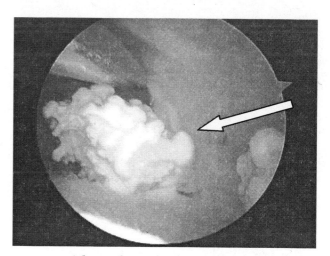

A large loose body in the knee.

In larger defects, local transfers of cartilage plugs from one location in the knee to another work well. The only downside is that the donor site may have more importance than your doctor realizes, and its surface will be compromised after taking the plugs. Newer synthetic scaffolds have been approved to fill these defects for this very reason. The synthetic

The removed loose body.

plugs have been used for small primary defects with good clinical results as well (this could be considered an FDA off label use. Even with good results in some surgeons' hands, it is not widely used at this time).

With even larger defects, frozen allograft (human donor tissue) has been used to cover the surface. Freezing preserves the cells in this donor cartilage. However, the survival of these cells is limited. It is also known that kissing lesions (defects on both the tibia and femur that touch) cannot be treated with two allografts. Still, large defects on the femur can be replaced with frozen grafts, yielding good results.

Lastly, tissue can be harvested at the time of a diagnostic arthroscopy; the cartilage cells are grown in a lab and later re-injected under a flap of periosteum (the thin membrane that covers bone). This is known as ACI, short for Autologous Chondrocyte Implantation. It is good for larger defects but requires at least two procedures, a time interval between them and a long rehabilitation period. MACI is a similar procedure available in Europe and Australia, where the cells are placed in a matrix before implantation. This makes locating and placing the cells easier and eliminates the need to harvest and sew the periosteum in place. New technologies are being developed all the time. Allograft cultured "neo" cartilage cells are being tested as a positive surface replacement. This technology also looks very promising. Hopefully these will be available in the U.S. soon.

Summary

In summary, small defects can be treated with the microfracture technique, but larger defects require grafting of some type (an autograft, allograft, cultured cartilage or newer synthetic bone substitute that acts as a scaffold and fills in over time with your own cells). The full healing process takes time for all repairs. If you have one of these repairs, you will use crutches and a brace to protect your knee from impact loading (this is usually the original cause of the defect). This allows the body to work on healing. To protect these repairs, many surgeons will use a knee immobilizer first that allows for a little weight bearing while using crutches. After the initial healing phase of four to six weeks, a knee-hinged brace is often used to continue protection of the healing cartilage for the first three to four months after surgery. A brace is needed for heavier activities for up to one year after surgery. Some patients with large defects and bowed legs may need a special unloader brace to allow a return to activity while the graft incorporates and the new cartilage grows.

Kneecap Pain and Dislocations
(Patella Pain, Chondromalacia, Subluxation and Dislocation)

The kneecap, or patella, has many important functions in the knee. It protects the front of the knee, it carries the forces from the thigh muscles to the tibia and it moves the patella tendon away from the center of the knee, improving the knee's mechanics. In addition, because of the kneecap's position in front of the center of knee rotation, it makes the muscles supporting the leg more efficient. As a result of this efficiency, the kneecap allows us to run, kick, jump, squat, balance our body weight on one leg while going down stairs and even makes it possible to get out of a chair. And, because of its importance in all of these functions, when it is damaged, dislocated or is tracking abnormally, we notice. When the kneecap is not doing its job, many people will feel anterior knee pain, pain with climbing stairs, squatting or kneeling and/or a sense of the knee giving way.

So what ails the mighty kneecap? The most common problem is simple wear of the undersurface. This is often known as chondromalacia patella. Chondromalacia is the loss of the articular cartilage's structural integrity. The underside of the patella is covered with articular cartilage which helps it glide in the trochlea (the special groove in the femur). The patella and the trochlea together are called the patellofemoral mechanism. Problems begin to occur with wear and tear, and the underlying cartilage begins to degenerate. The degeneration can occur from normal aging, because of the way the patella moves in the trochlear groove or excessive loading of the kneecap. The patient complains of pain on stairs, squatting, kneeling and running downhill more than up. There may be swelling, mal-tracking of the kneecap and a crackling sensation upon bending. Sometimes the sounds are loud and ocationally can remind people of a creaky door hinge. The problem is most often agrivated by overuse, being overweight, inappropriate weight training exercises, trauma, sports and mal-alignment or mal-tracking of the kneecap.

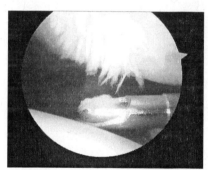

Chondromalacia patella with arthroscopic shaver removing loose elements.

When the cartilage under the kneecap starts to fail, changes are seen on its surface. In patients with chondromalacia patella, an exam through an arthroscope can reveal everything from a soft surface instead of its normal firm plastic-like consistency (grade one changes) to patches of advanced worn spots or cartilage loss down to bone (grade four changes). If the underlying issues are not addressed early on, arthritis and loss on knee function almost always follows.

Nonsurgical treatment

Orthotics can help decrease foot pronation and help unload the kneecap. Weight loss is also important when indicated, and activity modification can be a helpful part of the treatment as well. Other treatments include Glucosamine and Chondroitin. They are oral cartilage food supplements that are believed to reduce inflammation and support cartilage growth (see side bar on this page).

Patella subluxation or dislocation

Another cause of chondromalacia patella and source of pain may come from the tracking of the kneecap. As the knee bends, the patella is supposed to slide up and down the center of its groove. A normal patella should move relatively straight up and down the trochlea (its groove). However, the patella can slip out of place due to injury or congenital abnormalities of the shape of the knee. Sometimes the groove can be steep, sometimes very shallow. Sometimes bands of tissue holding your patella in place can become too tight on one side, causing the patella tendon to be aligned to the inner or outer side of the

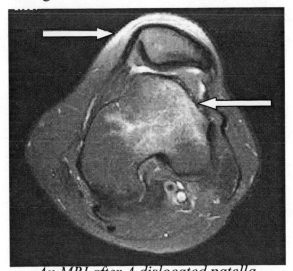

An MRI after A dislocated patella.
Note the bone bruise and the patella off center

When these problems occur, the initial treatment includes rest, ice and NSAIDs. Therapy for mal-tracking, including patella taping and VMO (Vastus Medilis Obliqueus) strengthening can help (see exercises in Appendix II). Patella braces can also be used, these benifit some people but, often there are mixed results.

knee. When the trochlea is shallow or the tendon is mal-aligned, the kneecap can jump over the edge of its groove. This is more commonly known as patella subluxation or dislocation. In subluxation or dislocation, the shift can weaken the soft tissue support for the patella in its groove. If it only partially slips out, it is called a subluxation. It could be minor and hard to see, or more serious and you may actually be able to see that it's in the wrong position just at first glance. If it completely slips out of its grove, it is called a dislocation. This can be worsened from trauma or a sports injury. If subluxation happens frequently, it will cause pain and disability.

Recurrent patella subluxation or dislocation gives rise to the sensation of the knee "giving way". Sometimes it's hard to describe the feeling since other knee problems (like ligament or cartilage tears) also cause the sensation of "giving way". The patella instability is also associated with pain along the outside or underneath the kneecap with squatting, bending or going up and down the stairs just like chondromalacia of the patella (as in the prior chapter). The location of the pain, along with a good physical examination, can help make the correct diagnosis.

After a dislocation, the bone on the patella and the edge of the femur may be bruised from the patella slipping over (see image: MRI of subluxed patella after an acute dislocation). In cases of more significant trauma, as in some collision related dislocations in sports, a piece of the bone and cartilage can be sheared off as the patella dislocates or comes back into place. If your Orthopaedic surgeon finds that the knee fluid shows a sign of blood and fat droplets, further study is required to rule out a fracture or loose fragment of cartilage in the knee joint. This would include an X-ray and possible an MRI. If an X-ray shows a lose chip of bone after a dislocation, it will need to be either removed or fixed, depending on the size of the piece. If an MRI shows a loose fragment, the treatment is the same as if an X-ray showed a loose fragment of bone. Sometimes only a piece of cartilage is broken off. In those cases, an X-ray may be normal since the loose piece can only be seen on an MRI.

Other patients have an anatomic divergence between the tibial tubercle and the center of the groove the kneecap sits in. In other words the attachment of the patella tend is not aligned with the grove increasing the likelihood of patella subluxation or dislocation. Measuring this distance, called the TT-TG distance (tibial tubercle–trochlear groove distance), helps in sorting the best treatment. When it is high (greater than 1.4 cm in females, 1.0 cm in males), there is increased mal-tracking and this may need to be addressed if the patella needs realignment.

Treatment

Physical therapy may be first prescribed to correct the patella alignment. Patella femoral pain associated with chondromalacia can also occur if the quadriceps is weak, which causes an imbalance in the joint. Since the quadriceps dictates the movement of the patella, the weakness can cause the patella to pull to one side more than the other. This puts more pressure on one side and can damage the cartilage over time.

Physical therapy, patella taping as described by Jenny McConnell (an Australian physiotherapist who invented the taping technique), orthotics and bracing are all aimed to correct the quadriceps muscle's pull on the patella.

If conservative measures fail, surgery may be required. There are two types of surgical procedures that can help fix these problems. The first is a Lateral Release. If the outer side is tight, painful and reproduces the patient's pain on palpation, this type of surgery may be helpful. This procedure releases the tight tissue near

tella realignment procedure, see diagram above). The tibial tubercle is the bony prominence below the patella. In this operation, the location where the tendon attaches to the tibial tubercle is moved forward and towards the inner side. It is then held in place by two screws while it heals, helping the patient become active again sooner. The tight lateral structures are also released in the procedure. The final goal of the surgery is to hold the patella in its normal groove, correcting the tendency for it to slide out of position to the lateral side.

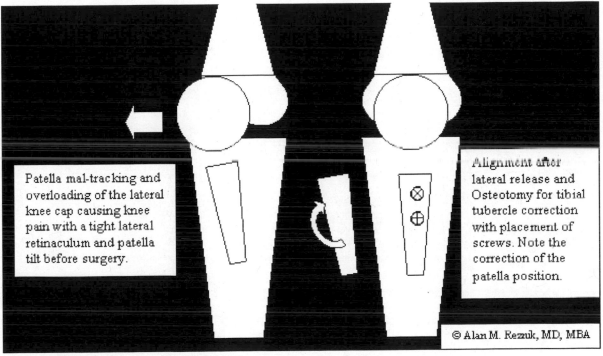

Patella mal-tracking and overloading of the lateral knee cap causing knee pain with a tight lateral retinaculum and patella tilt before surgery.

Alignment after lateral release and Osteotomy for tibial tubercle correction with placement of screws. Note the correction of the patella position.

the outer kneecap and helps to shift it to the medial side. It can be done arthroscopically, and a video of this procedure can be seen on YouTube (www.youtube.com/DrAReznik). In many cases, this removes enough pressure to solve the problem. However, some people have both tightness of the lateral side of the knee and mal-positioning of the tibial insertion of the patella tendon (an increased or abnormal TT-TG distance). The best way to control the patella's position is then with the Tibial Tubercle Transfer (TTT, or pa-

Loose Medial Patella

Lastly, many patients have torn the ligament that connects the inside of the kneecap to the inner, medial, side of the knee. This ligament is called the Medial Patella Femoral Ligament (the MPFL). It attaches to the upper third of the kneecap, and when it is torn, the kneecap can dislocate or subluxate laterally with ease.

Today, there is some discussion about the role of the medial structure's looseness, as opposed to the lateral tightness, as a primary cause of patella instability.

The surgery for medial looseness would tighten the natural structures on the medial side of the knee. Over the years, many authors have described moving the VMO over to help solve this problem. More recently, arthroscopic versions of tightening the medial side have also had good results. Now, more surgeons believe that tears of the ligament that holds the kneecap to the medial femur (MPFL) are a source of patella subluxation and the main problem after traumatic dislocations of the patella. The main reason this issue is more recognized and discussed than it was in the past is that many of the injuries to the MPFL are self-healing. At a recent national arthroscopy meeting, it was presented that 90% of these heal without treatment. In severe cases, less than 10% of the time, the structures can be completely torn. The medial side instability can be tested by pushing the patella laterally with the knee relaxed. If the patella glides over the lateral edge of the femur without resistance, the medial side may be a problem. In those cases, when the dislocating patella is also problematic, the MPFL can be reconstructed with a local or frozen tendon graft.

In rare cases of chronic recurrent patella dislocations, **all three problems** can be present at the same time. The patient has a tight lateral retinaculum, increased TT-TG distance and an incompetent MPFL (I have seen this only a few times in my practice). In these cases, a tibial tubercle transfer, lateral release and MPFL reconstruction may be required at the same time

to solve the patella instability problem.

In summary, when we factor in the mal-position or alignment associated with the tibial tubercle, MPFL laxity or tears, VMO laxity or weakness and tight lateral retinaculum, the options for correcting the kneecap problem increase a fair amount. These options include NSAIDs, physical therapy, patella taping, orthotics and surgery. The surgery can correct lateral tightness, medical looseness, bone mis-alignment or any combination of these issues at the same time.

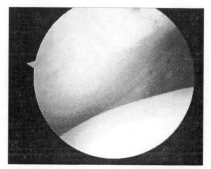

Arthroscopic view of a subluxed (tilted) patella.

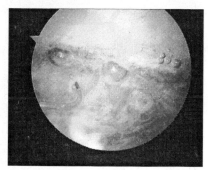

Arthroscopic lateral release using an electro-cautery.

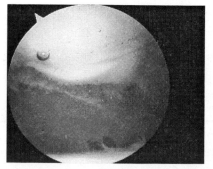

Patella tilt corrected.

Injury to the Knee Extensor Mechanism
Fracture of the Kneecap, Patella Tendon or Quads Tendon Rupture

Most people understand that the kneecap protects the knee; it helps the quadriceps work more efficiently, it enables us to go up and down stairs, get out of a chair, kick a ball, run and even jump. Many people understand what happens when the kneecap slips out of place or dislocates, but few understand what happens if the quads mechanism is broken by an injury.

Imagine, for a moment, a pulley and a rope. The pulley can break (the kneecap fractures), and the rope can break on either side of the pulley. On one side, the quadriceps attachment to the kneecap can tear (a quads rupture). On the other side, the patella tendon, the strong ligament that attaches the kneecap to the shinbone (tibia), can tear. All of these injuries remove your ability to balance on one leg and therefore make it almost impossible to walk. For all practical purposes, unless there are overwhelming medical reasons not to do so, these injuries need to be repaired.

Treatment

Patella fractures are failures of the bone. These can occur by direct trauma, as in a direct fall onto the knee, a car accident or a hyperflexion injury. The bone is more like cement than steel. It does well when force is applied in compression but not as well when it's stressed in

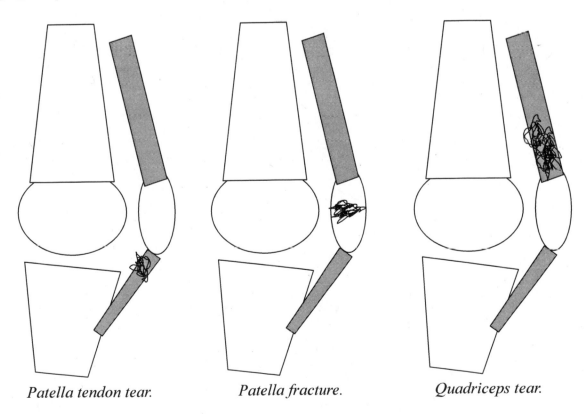

Patella tendon tear.   *Patella fracture.*   *Quadriceps tear.*

tension. As a result, a rapid load to the bending knee can cause the patella to split in two. When we add trauma to the surface, the fracture can be comminuted (split into multiple fragments). The treatments for kneecap fractures range from wiring the pieces together to complete excision of the kneecap. When there are very small pieces near the tip, the tip can be removed, and the patella tendon can be repaired directly to the bone. When there are two pieces, alignment of the cartilage surface and tension band wiring is the treatment of choice. Tension band wiring is the name we give to the method that converts the tension forces on the healing fracture to compression forces. As already stated, the bone likes compression more than tension. This repair method helps the bone press together as the muscles contract, allowing for earlier rehabilitation.

Patella tendon tears can be repaired to the bone directly. This works reasonably well in many cases; however, it applies all of the stress to the healing ligament and has the potential of failing to heal in a reasonable time. My personal preference is to use some type of tension relieving suture of wire through (or around) the patella to the tibial tubercle. In this way, the tendon repair can be protected as it heals and earlier motion can be accomplished safely.

Lastly, the quads tendon can rupture with a flexion injury or a fall. However, the tendon is very broad and strong, so rupture is not common. Some patients have a history of prior tendonitis, gout or inflammatory arthritis as predisposing conditions prior to this injury. These can be repaired directly to the bone using strong sutures and drill holes through the bone.

After all of these injuries and treatment, the patients should expect a prolonged period of protection in a long straight brace, stiffness and quadriceps weakness. In time, the tendon or the patella heals, range returns and strength follows slowly. Measured physical therapy is very important. Care must be taken not to over tighten the repair during the procedure and to protect it in the early phases of healing. To accomplish this, your surgeon may use any one of the tension relieving techniques as described above. He may combine this with fairly strict bracing, to be worn when not participating in closely supervised therapy with a licensed physical therapist. Patients should follow the rehab protocol and not pretend to "know better" than their surgeon or their therapist. As I have said before, Mother Nature has her own plans, and you cannot fool her.

Tibial Plateau Fractures

The knee has several weight bearing surfaces. The primary loads (weight) in the knee pass from the femur (thighbone) to the tibia (shinbone), with the curved surface of the femur resting on the relatively flat surface of the tibia. Like a mountain with a flat top, this flat surface is called the Tibial Plateau. This is a very sturdy surface, yet it is vulnerable to trauma and can break (fracture). The most common injuries result from a side blow to the knee. This can occur in sports, like skiing and football, or from trauma, like a fall or a car accident. The fracture below occurred when a large dog accidentally knocked its owner over from the side. The stress applied to the outer side of the knee can cause one of two injuries: rupture of the medial ligaments (medial collateral ligament sprain or tear) or collapse of the lateral plateau as seen here. You can imagine how the femur acts as a hammer as it hits the plateau in this type of injury.

There are many types of plateau fractures. These involve the outer (lateral) side, inner (medial) side or both sides (bi-condylar) of the plateau. If the surface is depressed or the sides of the bone are cracked, the plateau can no longer support the femur. This is made even worse if there is a ligament injury associated with the fracture. The unstable knee will be painful, unstable, swollen and often grossly deformed after the injury. The fracture can be detected by checking the medial and lateral stability of the knee, getting plain X-rays and performing CT scans (as in the CT generated image below in Fig 1). When there is instability during the exam or the fracture is significantly depressed (pushed down into the bone), it should be surgically repaired to preserve knee function.

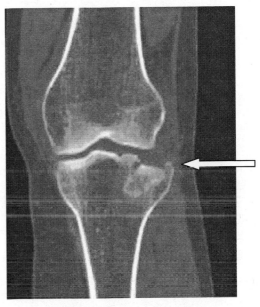

A Coronal CT Scan of a Fracture
This reconstructed image (made of many images formatted by the computer) shows a lateral compression fracture through the lateral tibial plateau. The goal of treatment is to restore the height of the depressed fragment seen here.

Treatment

Surgery is indicated when the surface is depressed or displaced significantly. Many years ago, up to one centimeter of displacement was accepted for surgery. Now, with CT scans and arthroscopic assisted techniques, no more than ½ a centimeter is accepted and, for some

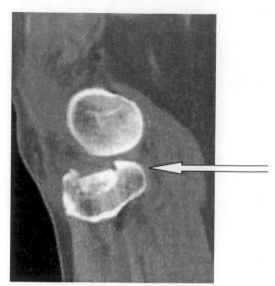

Side, or sagital, view of the same fracture.

surface at the time of the fracture. Many of the cartilage cells are killed in the initial blow to the knee, and that cannot be reversed.

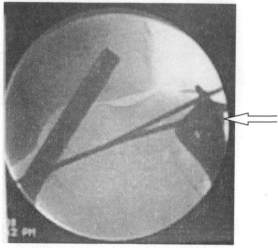

Intra-operative X-ray of the scope in place and a trial positioning of the plate with the bone elevated to its normal height.

doctors, even less than that. If the plateau is injured and the fracture is depressed, the femur will "fall" into the defect, and the knee will be unstable in the direction of the fracture. Walking on the fracture will worsen the condition. Once the plateau is fractured, the patient should be braced or splinted and placed on crutches. A CT scan is used to "see" the displacement, the number of fractures and the location of the pieces. In can also aid in surgical planning if surgery is necessary. At the time of surgery, an open or arthroscopically assisted method may be used to reduce the fracture and re-align the joint. Bone graft, screws and/or a plate with screws will also be used to support the surface. Newer plates with "locking" screws have improved the strength of these repairs, and pre-contoured "anatomic" plates have also improved our ability to get a good reduction of the fracture with a better restoration of the normal anatomy. The case below shows the use of a pre-contoured locking plate. Remember that even with a perfect reduction, ideal plate fixation and bone grafting, we cannot undo the crushing injury to the cartilage

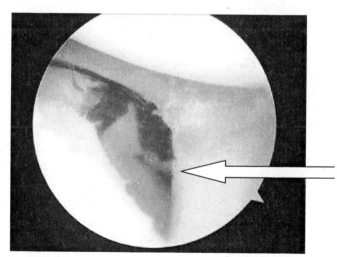

An arthroscopic view of the same fracture: Note how the depressed tibial plateau is no longer supporting the lateral meniscus. This explains the lateral knee instability.

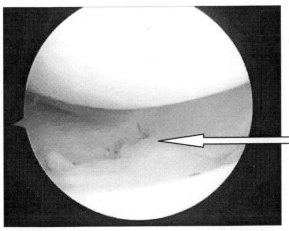

An arthroscopic view of the reduced fracture while the plate and pins are placed. The meniscus is now supported by the bone below.

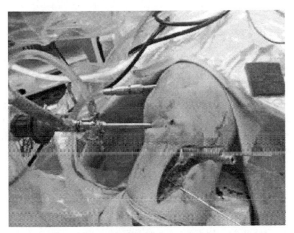

The plate is being placed in position, with the scope inside the knee checking the reduction. The fracture and plate are being held with temporary pins before the screws are placed.

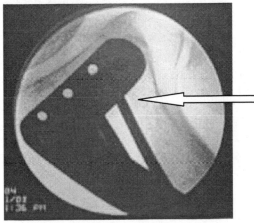

Intra-operative fluoroscopic X-rays: Side view of the plate in place.

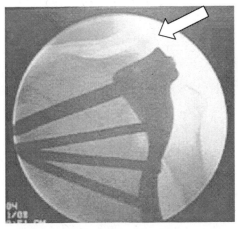

Final view of the fracture reduced, the bone graft in place and the plate affixed to the bone with the locking screws to fully stabilize the fracture.

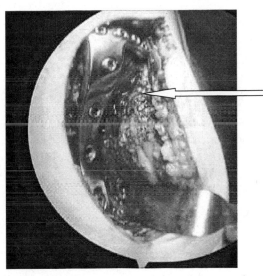

This is the final position of the plate in place with eight screws, as seen through the lateral "lazy S" incision. Four screws are holding the tibial plateau surface up; four are fixing the plate to the shaft of the tibia and the medial side of the knee.

Once the fracture is reduced and fixed in place with the plate and screws, the wound is closed, and the patient is placed in a straight knee brace to protect the knee. He or she may not fully weight-bear for at least 6 to 12 weeks post-operatively, depending on the nature of the fracture, depth of the defect, amount of bone graft used and the surgeon's assessment of the quality of the bone fixation.

Shoulder

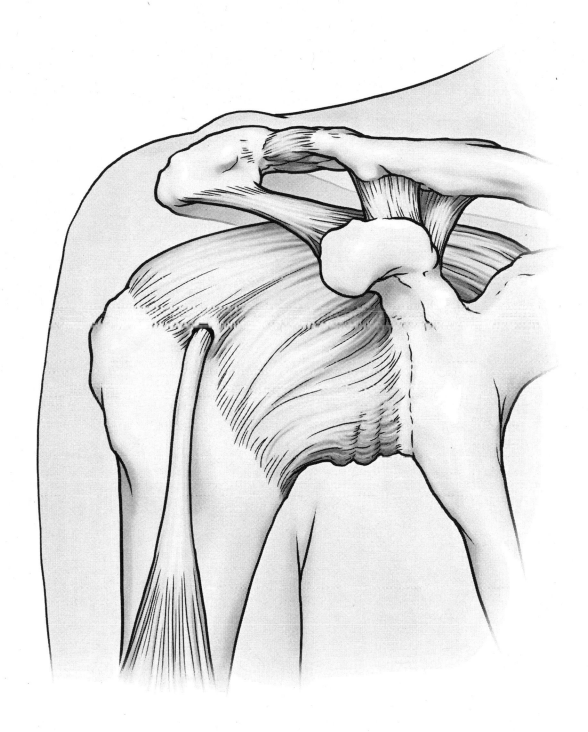

Frozen Shoulder (Adhesive Capsulitis)

Adhesive capsulitis, or frozen shoulder, can occur for many reasons. Most often, there is a smaller underlying shoulder injury or shoulder problem that causes the patient to decrease the use of the arm. In time, the patient barely notices the fact that his or her range of motion is slowly decreasing. The lining of the joint is, in fact, shrinking in size without the patient realizing it. At some point, activities of daily living become more difficult. Pain often increases, and eventually an orthopedic consult is requested. There is a strong association with diabetes in some populations. Diabetes itself seems to affect the lining of the shoulder, causing the lining to stiffen and shrink quickly after even the most minor insult or injury. The lining thickens with scar tissue and becomes less flexible. In some cases, all motion is lost. Moreover, a diabetic's frozen shoulder tends to be harder to resolve than a non-diabetic's frozen shoulder. On very rare occasions, a frozen shoulder can be the first sign of underlying diabetes and may be the only presenting symptom of early diabetes. Other patients who have skin hyper-sensitivity or have a history of forming keloids may also have an increased risk for a frozen shoulder (see side bar on keloids). Women are more susceptible than men. In addition, women are often seen earlier in the process than men. Because of the need to put a bra on, they notice the loss of internal rotation before men do.

Keloid Treatment

What is a keloid?

A keloid is a scar that doesn't know when to stop. When the skin is injured, cells grow back to fill in the gap. Somehow they "know" when the scar tissue is even with the contour of the skin, at which point they will stop multiplying. When the cells keep on reproducing, the result is what is called an overgrown (hypertrophic) scar, or a keloid.

Recommended Treatments

- Hydrocortisone cream every day to reduce inflammation, itch and inflammatory response. If itch persists, an antihistamine (such as Benadryl) may also be helpful.
- Silicone bandage overnight, such as Curad Scar Therapy.
- Continue treatment for at least two months.

What to avoid

- Sun exposure (Cover the scar when in the sun with a bandage or zinc oxide cream. Sunscreen or sunblock does not give enough protection for surgical scars).
- Scratching, rubbing or irritating the area.

Treatment

The **initial treatment** for a frozen shoulder is anti-inflammatory medications and therapy. If the shoulder has been frozen for a long time, a cortisone injection will help to "loosen things up" and help the therapist get things moving again. A Medrol Dosepak (a tapered dose of oral steroids) can also be helpful. A high percentage of patients improve with non-surgical methods. As long as they keep up with the home exercise program and anti-inflammatory medications as needed, the risk of recurrence is reduced. If a patient's capsulitis does not resolve or recurs, he or she may need a manipulation under general anesthesia, an arthroscopic lysis of the adhesions and/or a surgical capsular release. If there are underlying shoulder problems, they can be treated at the same time. For example, if large spurs or osteolysis of the clavicle are the primary causes of shoulder pain, they need manipulation and lysis of adhesions so that the frozen shoulder will not have a 'reason' to return. In all cases, initial treatment should consist of NSAIDs, therapy and injection. If a patient fails three or more months of treatment, surgery may be necessary.

The **success** of a surgical release depends on several factors. Once the manipulation is carried out, the most important goal after surgery is to prevent the scar or adhesions from reforming. Scar recurrence can be avoided by a very early and aggressive therapy program. Good pain control and anti-inflammatory medications are also important. To accomplish this, my patients are asked to start their exercises in the recovery room and start formal therapy a few days after surgery. At times, post-op anti-inflammatories are absolutely required to keep the inflammation in check. In difficult cases, an oral steroid may also be useful (Medrol Dosepak - a self tapering dose package).

Since many frozen shoulders are insidious in nature, the muscles have often been tight or contracted for some time. Many patients can feel this tightness and require both time and therapy to overcome the muscle shortening, even when the adhesions are fully released. The tight muscles and the body's own defense mechanisms can defeat even the best releases, so the close, watchful eye of a talented therapist can make all the difference. If a patient is doing well at first, but the motion seems to be deteriorating each week, medical management of the recurring inflammation is urgently required. Please tell your doctor if you feel like you are going backwards in recovery any time during the treatment. If this does happen, new medications may be tried, such as muscle relaxers (like Skelaxin or Flexoril), drugs to settle the nerve endings (like Elavil or Trazadone) and others up to and including oral or injected steroids. In some cases, there will be great improvement to a point, and the shoulder will be more functional but still not fully recovered. In those rare cases, after physical therapy and medications have failed, a second manipulation may be very helpful.

Remember, a frozen shoulder is best treated non-operatively first. Only those that have failed maximal medical treatment and therapy should consider surgery and only when they are prepared to work hard in therapy afterwards. But, don't wait too long! Many patients have symptoms for more than a year when I see them; they are fed up and tired of the constant limits to motion and pain with activity. They are therefore happy to comply with the therapy program but would have been better served if seen by an Orthopaedic Surgeon sooner.

When successfully treated, the patient with a frozen shoulder regains forward flexion or elevation first and then external rotation and internal rotation last. Men notice this when they are better but still cannot get their wallet out of their back pocket. Women notice this when they are having more motion and less pain but still cannot snap a bra behind their back.

Shoulder Instability and Dislocations

If you have seen it in the movies, you may have a vivid memory of a very well known actor popping his shoulder out and then putting it back in by slamming it against a locker room locker; gross! You may have seen contortionists put their arms in truly unreal positions or an escape artist get out of very tight chains by popping his or her shoulder out and then putting it back in. A few of you may even have friends who voluntarily shift their shoulders in and out of the socket. These have lasting visual effects because they are all abnormal motions of the shoulder and represent fundamental problems with the normal shoulder structure and function. In general, there are two main types of shoulder dislocations: those that result from traumatic first time dislocations and those related to a natural ligamentous laxity that a given person has as a genetic predisposition. Some dislocations can go back in place with ease, while others require emergency treatment with some type of anesthetic or sedation to get it back in place.

Here, we are going to look at these broad groups of patients in more detail so you can better understand the wide range of issues and treatments for these problems.

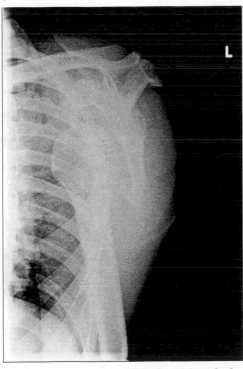

Scapula "Y" view of a shoulder dislocation. Note the humeral head (ball) is on front of the glenoid (socket)..

The shoulder joint has the widest range of motion of any joint in the body. This is true because the joint itself is positioned on the end of a very mobile stick (the collarbone, as discussed in other chapters), and the shoulder blade sits loosely over the ribs with no bone attachment to them. The socket is on the shoulder blade and is free to move widely with the motion of the blade, so its starting point varies greatly. The ball (humeral head) that sits in the socket is round, and the socket is very small and flat. Imagine a smooth golf ball on a slightly larger than normal tee. Unlike the hip, where the ball and socket are well mated and stable, the shoulder's ball and socket are not. So how does the shoulder stay in place?

The shoulder is referred to as subluxated or dislocated when the head of the humerus (the ball) has been forced partly or fully out of the glenoid cavity (the socket). To help stop this from being a daily problem for everyone, the shoulder joint is reinforced with ligaments and a rim of tissue that surrounds the glenoid cavity, called the glenoid labrum. That rim deepens the socket, making the lip wider and adding some stability. The ligaments add further stability, and the rotator cuff adds an active stability by pulling the shoulder into the socket when it contracts in a coordinated way with shoulder motion. The body of the scapula (shoulder blade) also moves to line up the forces on the shoulder in a direction perpendicular to its surface, reducing the risk of dislocation with heavy lifting, pushing or pulling.

Still, even with its protective labrum, rotator cuff muscles and alignment of the scapula with the loads applied, the shoulder can be dislocated more easily than any other joint. If any excessive force is applied to the arm in a direction close to that of the flat glenoid surface, just like a golf ball being hit off its tee, the shoulder may become dislocated. When this occurs, supporting ligaments of the shoulder may be torn, displaced or stretched out of shape. The labrum itself can be torn off the bone (the so called "Bankart" lesion). Furthermore, when the smooth cartilage surface of the head of the humerus slides over the labrum, it can become impinged against the sharp rim. This can fracture the lip of the socket or dent the back of the ball. When this happens, it is called a "boney Bankart" lesion on the lip side and a "Hill-Sachs" lesion, or defect, on the ball side. The boney Bankart and Hill-Sachs defect, by changing the normal anatomy, further destabilize the shoulder and can lead to recurrent dislocations. Naturally, the larger the dent in the ball or chip off the socket is, the easier it is for the shoulder to dislocate again. Almost always, at the same time as the ball is dented, the ligaments in front of the shoulder are avulsed, or torn off, the rim of the glenoid cavity. It is the combination of

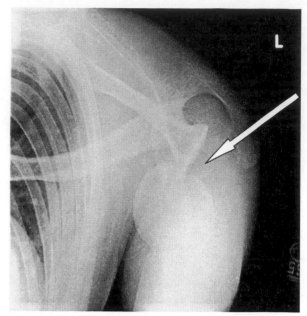

The glenoid impacted on the humeral head, causing a Hill-Sachs lesion.

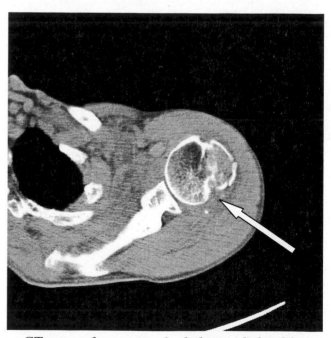

CT scan of a previously dislocated shoulder, post-reduction, showing a Hill-Sachs lesion.

the dent size and the amount of ligament damage that dictates the future instability of the shoulder. The same is true for a glenoid lip fracture; the ligaments come off with the lip piece and the larger the fracture, the more unstable the shoulder will be. In time, and with recurrent dislocations, more damage is done to the joint. This causes the patient to give up activities and, eventually, arthritis of the shoulder will develop.

What about the contortionist, and those people who are "double jointed"? They seem to be able to dislocate their shoulders at will, even without a traumatic event, torn cartilage, torn ligaments or damage to the bone. This is an interesting group of people. In general, they have naturally loose joints because the collagen that lines their joints and makes up all their ligaments is more flexible or elastic in nature. When checked, many of these patients have the ability to bend their fingers back more than normal people, have loose wrists, loose knees and elbows (they can be over straightened or hyperextended) and can touch their toes or the floor easily without bending their knees. This ability often runs in families and is considered to be something that your are genetically predisposed to having. Although it is natural for them to be able to subluxate their joints, once it starts happening on a regular basis, it may worsen and limit activity or cause pain. The treatment for these problems is not as straight forward as the traumatic type of dislocations and is a concern. Some of these patients have "learned" to dislocate their shoulders, and the muscle firing that causes the shoulder to come out cannot be unlearned too easily. In those cases, even the best treatment can be undone by the patient's own muscle action.

Treatment

The Bankart Procedure, Labral Repair and Gleno-Humeral Ligament Repair are all surgical techniques for the repair of the damage from a single joint dislocation or recurrent joint dislocations. In these procedures, the torn labrum with attached ligaments is reattached to the correct place in order to prevent further dislocations. With proper tightening of the lining, and if the dent is small enough, the Hill-Sachs lesion will not hit the rim with routine motion. The shoulder is thus more stable, and the chance of re-injury is greatly reduced by avoiding dent-

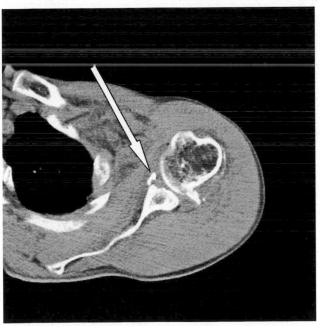

CT scan of a dislocated shoulder, showing a glenoid fracture and loose body.

to-rim contact. On rare occasions, the dent is so large that it must be grafted.

Traditionally, these were all done as open procedures. Now, with more modern techniques and better equipment, arthroscopic surgery has excellent outcomes with much less

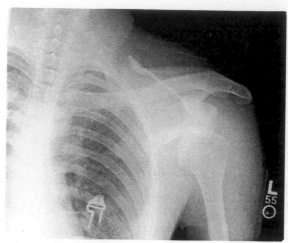

Front view of shoulder dislocation with humeral head (ball) inferior and anterior to the glenoid cavity (socket).

surgical trauma to the shoulder.

When there is a fracture of the glenoid, the lip fracture needs to be addressed at the same time as the ligaments are repaired. Very small fragments can be safely removed, and the labrum will be repaired over the defect. Slightly larger fragments can be reduced into place and are often repaired with the ligaments as a unit. Those involving more than 25% of the surface should be put back. Some of these may even require screw fixation. In some rare cases, the bone is badly damaged, and a bone block transfer is required (moving the coracoid process over and screwing it down into the defect with the short head of the biceps still attached and acting like a sling).

When the Hill-Sachs lesion is large, it too may require direct treatment. In some cases, moving the posterior rotator cuff into the defect helps reduce problems related to the loss of bone. In other cases, synthetic graft material or a piece of donor (frozen) humeral head may be needed to fill in the defect. A few surgeons will try to push the dent back out and graft behind it as another alternative. The choice is mostly

made by surgeon preference and is most often gauged by the size of the hole that needs to be filled. Most recently, using an anchor to secure part of the intra-spinatus tendon into the hole in the bone has been a very effective repair for medium sized defects.

Treatment for Loose Joints and Shoulder Subluxation

When the ligaments and capsule lining are stretched out of shape, they could cause instability and subluxation or recurrent dislocations. This can also occur traumatically or be a result of general ligament laxity (like the "double-jointed" people noted above). Many of the patients with this problem cannot do overhead work, overhead sports or throw anything. In addition to a good physical exam, an MRI arthrogram is helpful in making the diagnosis. If the ligaments alone are torn, loose or stretched, and the labrum is still attached, they can be repaired arthroscopically in a similar manner. To fix this problem, the loose capsule can be tightened at the same time as the ligaments are repaired. This is called a capsular shift procedure. Some patients with Bankart Lesions also have their ligaments stretched out when the injury occurs. When this is recognized, your surgeon can treat this at the same time as the repair of the labrum.

Special Cases:

Posterior Dislocations

Posterior dislocations are less common than anterior dislocations and therefore are often missed. They can occur traumatically, often from a severe trauma with other more serious or acute injuries. Occasionally, they are associated with seizures. In those cases, patients have no memory of the event (it is very normal after a seizure to lose memory of the event itself). When the cause is trauma to the shoulder, the arm is internally rotated and, to the untrained eye, in a seemingly normal position. It is only when someone tries rotating the arm externally that pain and a limit to motion is noted. Some of these trauma patients are unconscious and cannot help you make the diagnosis. A chest X-ray and even routine shoulder views are not at the correct angle to see the dislocation and are often read as normal. Only a true AP (Anterior Posterior) view of the shoulder (perpendicular to the plane of the scapula) will show these dislocations. As a result, many patients are diagnosed long after the injury and only by a much studied exam, special X-rays or a CT scan.

If seen and diagnosed early (the day of the injury), a close reduction is possible with sedation. Since these are often seen late, the treatment is almost always operative, and many times the defect in the bone is large and requires bone grafting.

HAGL Lesions

In even less common cases, the shoulder dislocates anteriorly, and the labrum is intact. This can happen if the ligaments pull off the humeral head side instead of the glenoid lip side, a Humeral Avulsion of the Gleno-humeral Ligaments (HAGL) lesion. In these cases, a repair may also be needed, but the arthroscopic repair is tricky and an open technique may have to be used.

Rotator Cuff Tears and Dislocations

There are patients that dislocate their shoulder without tearing the labrum or ligaments. Instead they tear the rotator cuff. These are generally in older patients (after age 40, there is a 60% chance of a rotator cuff tear with a traumatic dislocation of the shoulder), and repairing the cuff usually restores stability. Rarely, both the cuff and labrum are torn in the same injury. Again, these can be both repaired at the same time.

Torn Cartilage in the Shoulder: "SLAP" Tears

The shoulder is a ball and socket joint, similar to the hip; however, the socket of the shoulder joint (the glenoid) is extremely shallow. The labrum is a lip of cartilage that helps to form more of a cup for the end of the arm bone (humerus) to move within. As discussed in detail in the chapter on shoulder instability, this lip (or ring) of cartilage makes the shoulder joint much more stable, but still allows for a very wide range of movements. A SLAP tear is a specific injury to a part of the shoulder joint called the labrum. A very common labral injury is a tear that occurs on the top of the labrum, extending from the front to the back of the cartilage ring. This tear is known as a SLAP tear, meaning a **S**uperior **L**abral **A**nterior to **P**osterior tear. An injury in this area can be painful, cause a loss of range of motion and can sometimes be associated with a partial biceps tendon tear.

Tears of the labrum can be the result of a direct fall, a blow to the arm, a forceful lifting maneuver or repetitive trauma. Sometimes, the repetitive injury occurs in throwing athletes when the bump on the lateral aspect of the humerus (the attachment of the supraspinatus) hits the glenoid lip (the labral attachment and the biceps tendon anchor). The motion causes the labrum to peal back and come away from the bone. This eventually causes instability of the biceps tendon attachment. When this happens, symptoms can include a catching sensation and pain with movement, most typically overhead activities such as throwing. Patients often com-plain of difficulty sleeping. The tear decreases the stability of the joint which, when combined with lying in bed, causes the shoulder to drop slightly out of place. This in turn pulls on the muscles and ligaments, causing discomfort. If the biceps tendon is also involved, patients may complain of pain over the front of the shoulder while turning a screw driver or opening a door knob. The SLAP lesions are also associated with undersurface tears of the rotator cuff. Anterior SLAPs are associated with the undersurface tears of the anterior cuff, leading to the term, SLAC shoulder, or **S**uperior **L**abrum and **A**nterior **C**uff tears.

Superior Labral tears have been classified into many types. Type I is a tear of the inner margin and is not associated with any mechanical problem with the biceps attachment to the bone (the so called "Biceps Anchor"). These can be degenerative or traumatic and can be treated by debridement alone. Type II involves the biceps anchor, and treatment ranges from direct repair with sutures and anchors to moving the biceps attachment elsewhere with a repair (see the chapter on biceps tenodesis and tendonotomy). Type III is a bucket handle tear. The treatment is removal of the tear because the biceps attachment is not involved (just like a torn cartilage in the knee). Type IV is a bucket handle tear of the labrum with involvement of the biceps. This repair can include repair or relocations of the biceps, as well as repair or resection of the freely trapping labrum. Others

Pitchers and Other Throwing Athletes

Loss of Internal Rotation

Pitchers and other throwing athletes require special attention for several reasons. First and foremost, they require more motion and speed of arm rotation than any other athlete. When a baseball is thrown over 90 miles per hour, the shoulder rotation can be over 7000 degrees per second. To put it more simply, the arm reaches a speed that would cause it to spin around almost 20 times in one second. *Luckily the pitcher releases the ball in a very small fraction of a second, or we would really see some interesting shoulder problems!* That speed, in combination with the limits of natural motion, causes adaptations in a pitcher's shoulder that mere mortals don't have. First, extreme external rotation allows for the coiling up or the extreme cocking that starts the building energy for the pitching motion. Second, a compensatory loss of internal rotation leads to tightness of the posterior lining or capsule. Unfortunately, this combination of increased external rotation and a tight capsule leads to a condition called internal impingement. In these athletes, the top of the humeral head, rotator cuff and superior labrum hit each other, causing a SLAP lesion and an undersurface rotator cuff tear.

The initial treatment for internal impingement is therapy and a stretching program. This includes the so called "sleepers stretch". The athlete will lie on the right side with his or her shoulder abducted 90 degrees and the elbow at 90 degrees. Then he or she will self stretch by using the good arm to push the affected arm down toward the table in an internal rotation direction. About 90% of patients respond with increased internal rotation by gaining 10 degrees of motion within 10 days. Those that don't respond may need an arthroscopic posterior capsular release.

Core Control in Throwing Athletes

Ask a pitcher, soccer player, tennis player or even a hockey player to balance on one leg and squat part way down without losing his or her balance. Have him stand on one leg and try to throw or catch a heavy workout ball. See if he can stand on one leg and bend his opposite hip and knee to 90 degrees, and see if he can straighten his knee or kick from that position. These are all simple balance tests that many of the "high" level high school and college athletes frequently fail in my office. They too often lack good hip and core control. Newton taught us (remember any high school physics?) that for every action there is an equal and opposite reaction. So if I could throw a ball at 90 miles per hour or hit a serve at 125 miles per hour, where does the "equal and opposite" reaction force come from?

The simple answer is the arm. But then where does the force in the arm come from? It has to be the body. And the body's force, where does it come from? Now, the final answer is the ground. So at the end of the "force chain" is the ground we stand on, and our legs control how that force is transferred to the ground. The next concern we therefore have for a throwing athlete is how well he or she transfers force to the ground. This requires great balance. So what happens to the athletes who fail my three tests of balance noted above? They have overuse injuries to the shoulder and elbow. Why? They cannot smoothly transfer force to the ground. Their legs are not transferring weight efficiently. So they try to overcompensate by using the arm muscles to force speed into the pitching motion. When is this frequently seen? In my practice, I see this mostly at the end of a growth spurt and with an increase in the competitive level of play. For example, when a junior high pitcher has rapid growth, when a high school student moves to varsity or when a high school player goes to college level play. The treatment for this issue is balance training and core strengthening exercises and an evaluation by a pitching coach to be sure that the pitcher's form will allow for good weight transfer during the throwing motion.

have described more types of tears, including combined lesions that are from shoulder instability, but they are beyond the scope of this discussion.

Treatment of SLAP tears

Very few patients with SLAP tear injuries return to full capability without surgical intervention. In most cases, surgery is required to repair the tear and reattach the labrum to the glenoid. Dr. Reznik does this repair through the arthroscope with sutures and tiny absorbable anchors. The goal is to repair the tear and restore normal function in a minimally invasive way and as an outpatient procedure. When a SLAP tear is associated with significant damage to the biceps tendon, a biceps tenodesis (transferring the tendon to a new location and removing its attachment from the superior labrum) may be necessary to correct the shoulder problem.

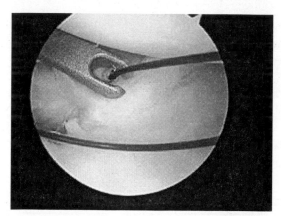

Lead suture being passed through a torn ligament/labrum.

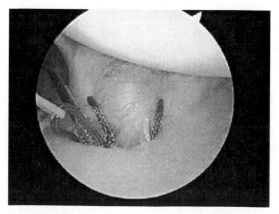

Permanent suture with anchor positioned in the bone in the glenoid (socket).

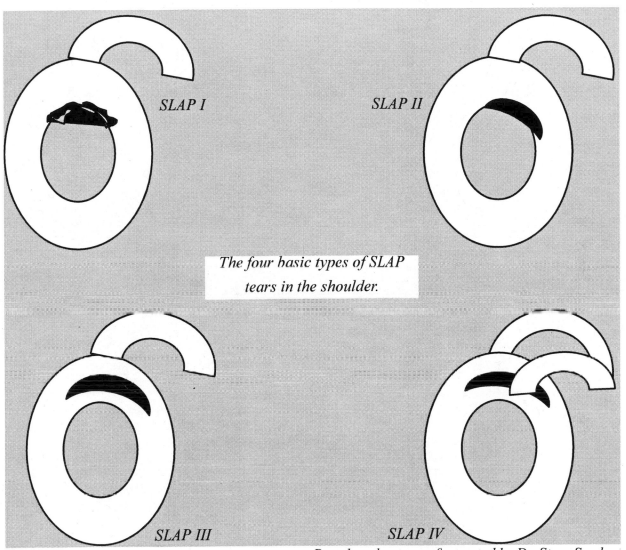

SLAP I

SLAP II

The four basic types of SLAP
tears in the shoulder.

SLAP III

SLAP IV

Based on the system first noted by Dr. Steve Snyder

SLAP types: I - torn edge, biceps anchor completely intact. II - torn with at least a partial detachment of the biceps ancho. III - bucket handle tear, the inner segment is free to trap in the shoulder joint. IV - labral tear with extension into the biceps tendon itself.

Rotator Cuff Tears

The tendons and associated muscles that cover the head of the humerus (the ball of the shoulder's ball and socket joint) are better know as the "rotator cuff". It is made up of tendons from four different muscles: the subscapularis, infraspinatus, supraspinatus and teres minor. All of these muscles begin on the shoulder blade in their own separate locations as broad muscle attachments to bone. They then narrow down to form tendons that attach at all sides of the humeral head. The four tendons directly control rotation of the ball in its socket and help keep the ball in place. The tendons grouped together form a strong "cuff" around the head, hence the name, "rotator cuff."

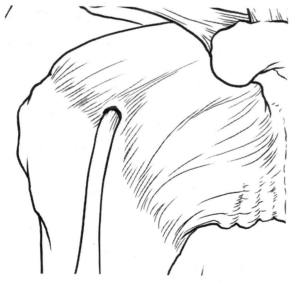

The "cuff" surrounds the joint, covering the joint capsule, the ligaments that stabilize the shoulder and the glenoid labrum, or "lip". As discussed in the chapter on shoulder dislocations, the labrum and ligaments help keep the very round humeral head on the very flat gle-

noid. The cuff serves to help these structures by supporting and strengthening the shoulder joint. With its muscles, the cuff controls the rotation of the humerus in the glenoid cavity (socket of the shoulder). By controlling rotation, it also helps to keep the ball in the socket. It initiates most shoulder movements and helps to keep the ball centered in the socket when the larger muscles are acting on the upper arm. This centering action is more important than it may seem. By keeping the ball centered in the socket, the cuff not only prevents dislocation, but also keeps the center of rotation of the shoulder in its place, helping the deltoid be more effective in lifting the arm. When the cuff is not working because of weakness, damage (torn tendon) or compression of the nerves that control it (like the suprascapular nerve, see the section on special conditions involving the rotator cuff), the ball does not stay centered in the socket. When the deltoid contracts and the ball is out of place, it will slide on the glenoid surface and the muscle force will pull the ball straight up instead of rotating it in its socket. The upward movement is only stopped when the ball hits the undersurface of the acromion. This cause both pain and further loss of motion. If the cuff is inflamed for any reason and weak, the impingement of its tendons against the bone above can be a cause of further pain and weakness. Some feel that the constant rubbing of the cuff tendon on the bone above it is one of the ways rotator cuff tears start to happen.

It makes sense that rotator cuff tears are often seen in people who are over 40 years old who do repetitive shoulder motions, overhead work, heavy labor, overhead sports or regular weight training. Sometimes patients have ongoing shoulder impingement and then an acute event after lifting something or falling. They may have had some chronic changes in the cuff, a thin cuff or a partial tear, then traumatically ruptured the balance of the tissue. Patients with

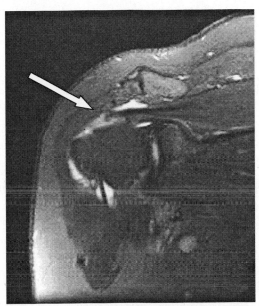

A small rotator cuff tear.

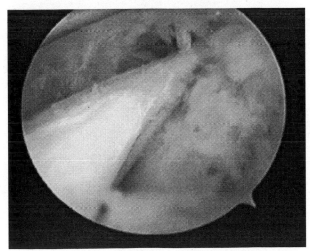

Arthroscopic view of the tear.

gout or an inflammatory arthritis also have tendons that are weakened by their disease and can tear their rotator cuff after a minor trauma or lift. Ongoing impingement and/or inflammatory weakening of the tendon are not prerequisites for a rotator cuff tear. Healthy tendons can also be torn at any time with enough force applied in the right direction. Tears can also occur in younger patients following acute trauma or a sports injury.

Tears can range from small and slightly open to large and completely retracted. They can be full thickness or partial thickness. The full thickness tears can be curved, or "C" shaped, "L" shaped, "V" shaped or "U" shaped (see figure on following page). The partial thickness tear can be within the tendon itself, on the upper or lower surface. Sometimes partial tears are associated with calcium deposits that can be seen on a plain X-ray of the shoulder. This is called calcified tendonitis. Tears can also be associated with a biceps tendon tear. If the biceps tears spontaneously, a good number of patients will have a rotator cuff tear too (up to 50% of them).

How do you diagnose a rotator cuff tear?

If you have a tear, you will experience loss of motion, weakness and pain. You could also have night pain and pain with certain arm motions, like overhead activity. The results of these symptoms are loss of sleep and the inability to lift common items (like a bottle of milk).

To make the diagnosis, your doctor will take your history and perform a physical exam. When taking the history of the problem, most of

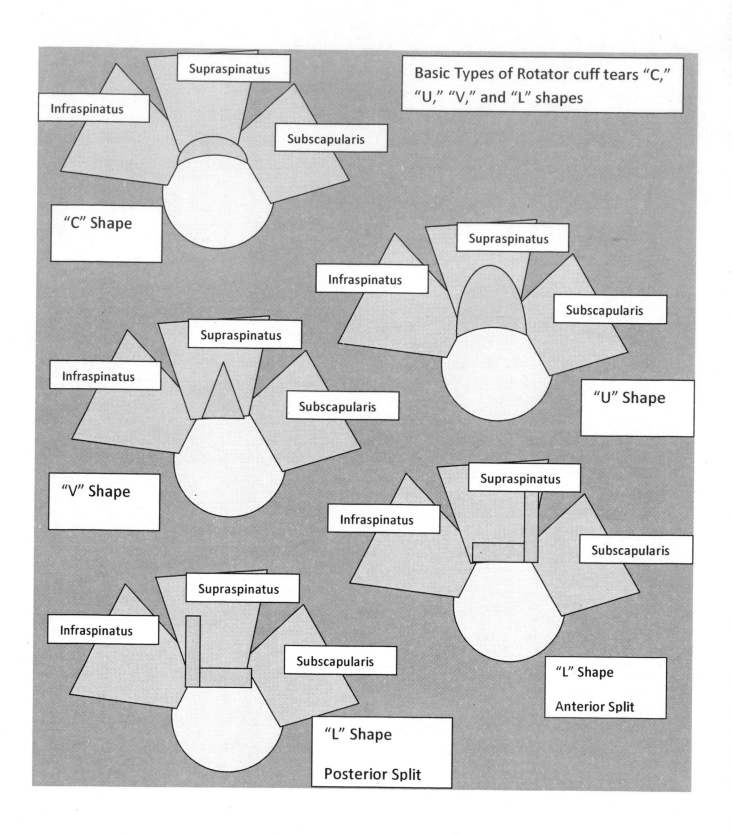

Basic Types of Rotator cuff tears "C," "U," "V," and "L" shapes

Supraspinatus

Infraspinatus

Subscapularis

"C" Shape

Supraspinatus

Infraspinatus

Subscapularis

"U" Shape

Supraspinatus

Infraspinatus

Subscapularis

"V" Shape

Supraspinatus

Infraspinatus

Subscapularis

"L" Shape

Anterior Split

Supraspinatus

Infraspinatus

Subscapularis

"L" Shape

Posterior Split

my patients say that they cannot put plates on a high shelf in the kitchen, take a gallon of milk out of the refrigerator with their arm straight or reach into the back of the car. Many say they drop things and most have trouble sleeping because of night pain. The exam almost always demonstrates weakness in elevation and impingement signs. Your doctor may give you a drop arm test. In this test, your doctor will lift

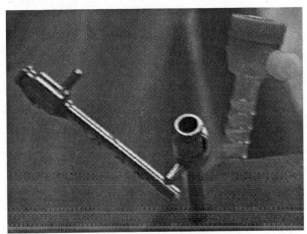

Arthroscopic repair.

your arm to 90 degrees away from your side and ask you to lower it slowly. If you can't do this, or with the slightest resistance the arm drops to your side, it indicates rotator cuff pathology. Some patients have external rotation weakness with their arm at their side, but not all patients do. Sometimes a plain X-ray will show a calcium deposit in the tendon or the humeral head riding higher in the socket than normal. A tear can be confirmed with an MRI.

Treatment options

A tear (mostly tendon swelling and partial tears) can be treated non-surgically through rest and a modification of activities. NSAIDs should be used to decrease the tendon swelling.

Physical therapy will also strengthen the other shoulder muscles that can help compensate for a tear, and, for patients who are not the best surgical candidates, this may be their only option.

If a patient fails conservative treatment methods, there is a full thickness tear with loss of use of the arm and/or the tear involves more than half the thickness of the tendon and remains painful, surgical repair is necessary. The type of repair depends on the size, shape and location of the tear. The patient's activity level, bone structure and the cause of the tear are also deciding factors in the correct treatment method. A partial tear (less than 50% of the fiber's thickness is torn) may require only a trimming or smoothing procedure (this is called "debridement"). Removing thickened bursal tissues or calcium deposits may also help. If there are bone spurs impinging on the tendon, they can be a source of pain and can also be removed.

A tear that is not at the attachment to the bone or within the substance of the tendon

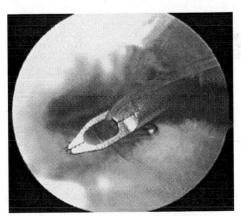

Rotator cuff suture passer.

can be repaired by suturing the two sides of the tendon together. If the tendon is torn from its insertion on the tuberosity of the humerus, it can be repaired directly to the bone using tiny suture anchors or bone tunnels through the bone. The

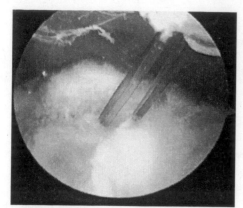

Rotator cuff suture in bone.

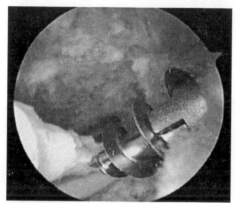

Rotator cuff anchor.

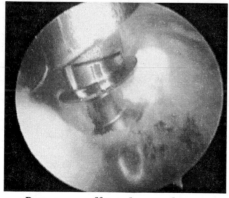

Rotator cuff anchor in bone.

arthroscopic cannula (a small tube that's inserted into the body). The cuff can then be cleared of scar tissue and debris. The inflamed bursas can be taken out. The bone can also be roughened to make a nice bleeding bed for the tendon to "heal" to. Once the repair site is ready, the tendon is mobilized, and "U", "V" and "L" shaped tears are converted to "C" shapes by side-to-side tendon stitches. The suture anchors are then introduced and placed in the bone. With the suture firmly in the bone, they can be passed through the tendon and tied in place. Depending on the size of the tear, repeating those steps multiple times will complete the repair.

Arthroscopic techniques are technically demanding, and patients should rely on their surgeon's judgment as to what the best procedure for their own individual problem will be. In experienced hands, complications are rare in routine shoulder arthroscopy. The complications would include stiffness and failure of the repair to heal. However, in general, the complication rates for arthroscopic repair are low. In my own experience, the risk of infection in shoulder arthroscopy is less than 1 in 2000. Also in my experience, the use of a scalene block to help with post-op pain control and the use of a pain buster pump for longer term post-op pain

surgery can be done as an open procedure (by taking the larger muscles out of the way and repairing the tendon directly), through a "mini open" approach or through an arthroscope. My preference is the arthroscopic method, although all the other techniques are widely used.

In the arthroscopic procedure, the rotator cuff is seen through a fiber optic scope inside an

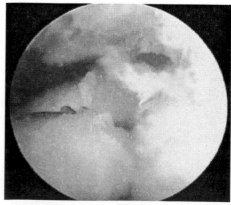

Rotator cuff repaired.

relief have higher complication rates than when the procedure is done in an out-patient setting, using an arthroscopic technique with a light general anesthesia and using local anesthetic at the surgical site to aid in post-operative pain control.

Special conditions of the Rotator Cuff:

Irreparable tendons

Sometimes the tendon will not be repairable. In these cases, special techniques to gain a partial repair are possible, synthetic grafting techniques have been tried and transfers of other muscles for those with very little function have been used. Standard shoulder replacements do not work well in this setting. A newer reverse shoulder replacement (the ball is on the socket side and the socket is on the ball side) may help the patients with complete irreparable tears and very low demands to regain good function, but they are not truly indicated in patients who are significantly overweight and require their arms to move about. For example, people that are crutch, walker or wheelchair dependent or use their arms to help them get up and out of a chair are at risk of loosening the prosthesis quickly with their normal daily activities. Remember, the indications for these procedures are limited.

In a good number of cases, mobilizing the tendons and a side-to-side repair (a margin convergence technique) along with removal of painful bone spurs and physical therapy does improve function. These patients are by no means fully functional, even with a good partial repair, but many are satisfied with the improvements.

Nerve issues

Some patients have nerve problems that cause shoulder pain and loss of function. The nerves can be pinched at the neck, as in a herniated disc. Surgery on the shoulder does not improve them. Some patients have more than one problem, like a rotator cuff tear and a herniated disc. Most will try to fix the tear first, since the surgery carries less risk. Again, if the disc is causing shoulder or arm pain, patients will not improve without it being treated directly.

Ganglion Cysts and Suprascapular Nerve Entrapment

In some glenoid labrum tears (see SLAP Tears) there are cysts that form above the socket near the exit for the suprascapular nerve. They press on the nerve, which can cause pain and weakness as if the patient has a tear, but there is no tear seen on an MRI. Decompressing the nerve by removal of the cyst and repairing the SLAP lesion will often solve this problem. The suprascapular nerve can also be compressed at the sphenoid notch, near a sharp turn it takes before it reaches the muscle. There is a band, or ligament, that crosses the nerve and can get tight. Releasing the band and freeing the nerve frequently solves this unique problem. Trying to repair or do a simple decompression of the space (an acromioplasty, as discussed in the next section) with an intact tendon is not helpful in these situations.

Shoulder Pain without Tears of the Rotator Cuff
Shoulder Impingement, AC Joint Pain and Pinched Nerves

The shoulder can be weakened or hurt from a number of conditions. A few have been discussed already. These include rotator cuff tears, subluxations, dislocations, labral tears, SLAP lesions and arthritic conditions of the ball and socket joint. Overhead use, trauma and repetitive motion can also cause pain when the rotator cuff rubs against the bone above it. In these cases, the tendon can swell or even be partially torn. The tendon can hit the front or side edge of the acromion or the undersurface of the connection between the acromion and the clavicle (the AC joint). Some people are born with a flat acromion and others with a curved or hooked acromion. These are known as type I, a flat acromion, type II, a mild hook, and type III, a significantly curved undersurface of the acromion (figures 1, 2 and 3). The patients with types two and three can be predisposed to the condition of sub-acromial impingement. In addition, the ligament that attaches the coracoid process to the acromion (it helps to distribute the load of the short head of the biceps and cover the front part of the rotator cuff) can calcify with stress in time. The hardened ligament can also rub against the rotator cuff, causing impingement type pain.

Impingement syndrome and acromioplasty

If someone has "impingement" from an overhanging acromion, the tendon is being constantly rubbed by the bone edge or

Figure 1: Type I Acromion

Figure 2: Type II Acromion

Figure 3: Type III Acromion

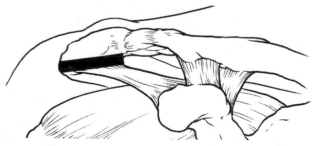

Bone removed for an acromioplasty.

spur. This usually presents itself as pain with overhead activity, occasional weakness with some motions and difficulty with lifting objects away from the body. The simple question is: can removing the overhang stop the rubbing? Similarly, can removing the calcified ligaments that often go with this disorder also help stop the pain? Moreover, if someone is born with a hooked acromion (type III), can we turn it into a flat acromion (type I)? Luckily the answer is yes to both of these questions. So, when impingement occurs and the pain fails to improve with non-operative treatment like NSAIDs, injections and therapy, arthroscopic surgery may be indicated. A famous shoulder surgeon, Dr. Neer from New York, popularized this as an open procedure many years ago. The procedure is called an acromioplasty. Approximately 20 years ago, the concept of doing this with an arthroscope started to gain acceptance in very small circles of surgeons, and about 15 years ago, one of my students reviewed many papers (one of the first Meta analyses reviews ever done) and showed that it was a better way to go. Today, arthroscopic acromioplasty has come into its own as a standard treatment for impingement syndrome. In the procedure, the overhanging part of the acromion is removed to convert the curved or hooked acromion to a flat one. If the ligament is calcified, it can be removed with the calcium deposits. If the ligament is tight, it can be pushed back (recessed) or cut (released). Also, during the arthroscopic procedure, the rotator tendons are checked for a tear. If the cuff is torn, the tears are treated as noted in the chapter on rotator cuff tears.

AC joint pain and Mumford procedure

In some cases, the end of the clavicle is also worn down, the AC joint is degenerated or arthritic and large spurs form at the edges. Some of the spurs stick down toward the rotator cuff and impinge on its surface. This can be painful for two reasons. First, the joint itself can hurt. Second, the spurs can impinge, rub or even tear the rotator cuff. In these cases, when conservative non-operative measures fail, removal of the spurs and resection of the diseased AC joint is the treatment of choice. This can very effectively remove the cause of the tendon rubbing and arthritic joint pain. This resection of the end of the clavicle is also known as a Mumford procedure.

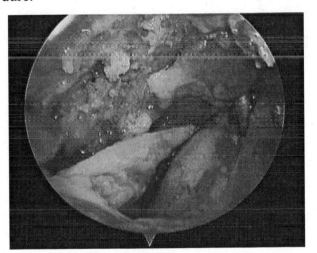

Rotator cuff tear, AC joint cyst.

Shoulder pain, or is it?

Some patients have similar symptoms, yet no true impingement signs are present on exam. The pain is referred down the arm to the elbow and/or hand. This means that the pain starts in one location, but is felt, or "referred", to another. This is just like when someone says their back hurts when, in fact, they have a kidney stone. They just don't know exactly where their kidney is. The same is true for the inside of the shoulder. When there is burning pain, spasm of the trapezius (the muscles around the lower neck), muscle weakness or neck stiffness, we sometimes need to look away from the painful shoulder for the real source of the pain. In many cases, these patients have a pinched nerve. The nerve can be trapped beneath the muscles that are tight, beneath an arthritic spur in the neck, a herniated disc in the neck or, in very rare instances, by an anatomic variant like an extra rib. Frequently, in these cases, the finding of burning, tingling or numbness may be key to making the diagnosis. Sometimes, the only tip off to this problem will be no pain relief from a shoulder injection. To make things more complicated, in some patients both situations occur at the same time. This can be frustrating since the patient responds only partly to treatment of the shoulder pain and at the same time, only partly to treatment of the neck problem. Sometimes the spasms from chronic shoulder pain cause the neck pain to worsen, and treatment of the shoulder can help with the pain. This is truly an individual situation for each patient, and each case is different in this respect. If the shoulder treatments, including therapy, NSAIDs and injection, fail, and X-rays and/or MRI of the shoulder yield little helpful information, a thorough nerve exam should be done and neck X-rays should be taken. In my practice, I like to add flexion and extension views to the series of films to look for abnormal motion of the vertebrate or lack of motion as a sign of significant muscle spasm or disc abnormality. If weakness and pain persist after non-operative treatment for the neck, an MRI of the neck is indicated. If there is a disc herniation, a consultation with a spine specialist may be necessary.

Wear of the collarbone

In patients who are extremely active, lift weights overhead or work in heavy labor, the end of the collarbone can develop internal swelling and microscopic fractures. Just like stress fractures in military recruits in boot camp, these can be painful. Rest, ice and NSAIDs help. Increased calcium intake and vitamin D can give your body the tools to heal the micro fracture, but continued lifting and repetitive stress to the bone under attack can cause the undersurface to fail under the pressure. When this occurs, cysts form in the bone, the surface of the AC joint can become arthritic and the distal clavicle end can widen, causing impingement. When the bone break down can be seen on plain X-rays or cysts can be seen on an MRI in the end of the bone near the AC joint, this condition is called osteolysis of the clavicle. Weight lifting overhead is painful and extremely limited. If rest has not helped and the bone is eroded, a resection of the distal clavicle (a Mumford procedure, see drawing below) works well in reducing pain and improving function. Again, this can be done either open or arthroscopically.

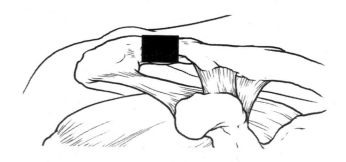

Injuries to the Biceps Tendon
Biceps Tendonitis, Partial Biceps Tear, Biceps Subluxation and Biceps Rupture

The biceps muscle starts at the elbow, passes up the arm and splits into two tendons, or "heads". The shorter tendon ends at the coracoid process of the "shoulder blade" (the scapula), and the longer one enters the shoulder joint. There, the longer end (the "long head") attaches to the top of the socket (the glenoid) at a cartilaginous lip that covers the edge of the socket (the labrum). The long head of the biceps can be injured by repetitive motion, local trauma, rapid extension of the arm, force applied while trying to actively flex the elbow or during a fracture or dislocation of the shoulder. It can be associated with rotator cuff tears and can subluxate with subscapularis tears. Its true function in the shoulder joint is heavily debated in the orthopedic and sports medicine community. At the elbow it acts, together with the short head of the biceps, in flexion of the elbow and with supination of the hand (clockwise rotation of the right hand and counter-clockwise on the left).

Biceps tendonitis is the most common problem seen in the long head of the biceps. It can often be treated with anti-inflammatories, ice and rest. In chronic cases, injection or therapy may be needed. Occasionally, there is a structural issue, and tenolysis (release of the tendon sheath), an arthroscopic decompression of the shoulder or a tendonotomy (release of the tendon itself) may be required.

A biceps tendon injury typically involves a partial or complete tear of the longer tendon. It is more common after the age of forty. Many times it is associated with an acute injury or a painful pop. Then, the muscle attached to the long head tends to "ball up" further down the arm. The "short head" almost always remains intact. Prior to the injury, some patients often have a long history of inflammation of the tendon (chronic biceps tendonitis). When that occurs, it is usually due to years of wear and tear on the shoulder. It is often associ-

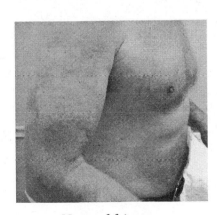

Normal biceps.

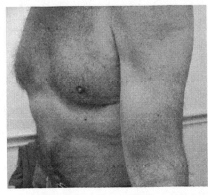

Ruptured biceps tendon, arm in relaxed position.

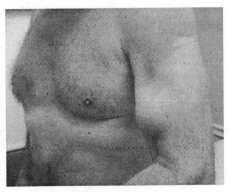

Ruptured biceps, arm flexed with the classic "Popeye deformity."

ated with repetitive overhead lifting, chronic inflammatory tendonitis and a heavy lifting injury, or repetitive work trauma. In other cases, a more severe, sudden traumatic injury is the cause. This is more common in younger patients but can occur at any age. A traumatic torn biceps sometimes occurs during heavy weightlifting or from actions that cause a sudden load on the upper arm, such as a hard fall with the arm outstretched during competitive sports. Forced extension of the elbow against resistance or a fall in a position that forces the tendon to trap between the humeral head (ball of the shoulder) and the 'sharper' bone edges of the scapula or acromion can also cause a tear or rupture.

Please note: Tears of the biceps tendon at the elbow are a completely different problem. Both heads of the biceps muscle join and attach in one common location on the proximal radius. Together, they are a major flexor of the elbow and these tears should, in general, always be repaired. This discussion is focused on injuries to the long head of the biceps.

Treatment options

Many partial and even complete tears can be treated without surgery. A well performed physical exam by an orthopedic surgeon and an X-ray of the shoulder are often the best ways to see what treatment is most appropriate. About 50% of long head of the biceps tendon ruptures are associated with rotator cuff tears (mostly supraspinatus tears). Subluxation of the long head is associated with a subscapularis tear. If there are signs and symptoms associated with a rotator cuff tear found on examination,

further testing may be needed. When a patient has significant symptoms, an MRI is frequently required to make the diagnosis of other shoulder problems associated with a biceps rupture.

Surgery is often reserved for patients with evidence of other concomitant shoulder problems. When the long head of the biceps is completely torn, the acute soreness will resolve in weeks. Some patients actually feel better than before the injury. If the muscle itself is painful with activity, the shoulder needs to be examined. If patients have weakness and pain with supination of the hand (clockwise rotation of the right hand and counter clockwise on the left) after failing conservative measures, a biceps tenodesis may be required. Biceps ruptures are also frequently associated with bone spurs near the tendons path into the shoulder joint. When these are painful they should be removed. If an MRI confirms the rotator cuff is torn, it should be repaired at the same time.

Special considerations

When the biceps is subluxated (out of its normal grove) and the subscapularis is torn, releasing the tendon may be necessary to protect the subscapularis repair. When there is a tear of the superior labrum (a cartilage "lip" on the socket of the shoulder joint) that involves the attachment of the biceps, repair of the labrum (lip) is needed, and a release of the tendon may also help resolve the symptoms (this is a newer concept and there is no clear agreement on the best treatment at this time, your surgeon will have to make a judgment based on the findings at the time of surgery).

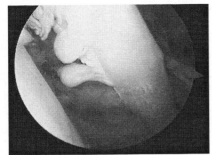

Massive tear and biceps tenodesis.

When needed, a **Biceps tenodesis** is a surgical procedure that anchors the ruptured end of the biceps tendon to the upper end of the humerus. Dr. Reznik performs arthroscopic evaluation of the shoulder to check for other related injuries to the shoulder first. Once any rotator cuff issues are treated, if a tenodesis is needed, it is done through small incisions over the front of the humerus. Depending on the length and condition of the tendon, the location of the tenodesis will vary. Newer arthroscopic tendon transfers are also possible in some cases. The type of procedure will depend on your anatomy and the problem found at the time of surgery. When tenodesis is performed, the tendon itself can be fixed in place with a special absorbable screw, sutures anchors or a combination of these methods. The surgery is done on an outpatient basis with the goal of decreasing pain with activity and improving overall function in the affected arm.

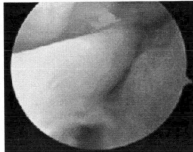

Figure 1: An Intra-articular view (inside the ball and socket of the shoulder) of a normal biceps tendon.

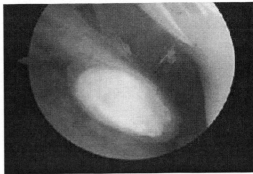

Figure 2: An inside view of a torn biceps tendon. See the torn end facing the camera head on.

Figure 3: The tendon delivered out of a small incision.

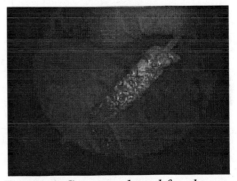

Figure 4: Sutures placed for the repair.

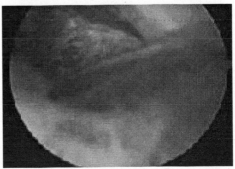

Fig 5: Tendon reattached to the bone in a new location.

Biceps Tenotomy

In some cases, the biceps is partially torn and painful. The tendon is swollen, worn, frayed or inflamed in its groove. Forward flexion of the arm, supination of the hand and pressing on (palpating) the bicepital groove is painful. If the non-dominant arm is involved, the patient has a low demand occupation and the shape of the muscle (cosmetic appearance) is not a concern, a tenotomy (a release of the tendon) can be an excellent option with good pain relief and a shorter recovery time than a tenodesis.

The AC Joint

The AC joint, the connection between the collar bone and the shoulder blade, is one of the most commonly injured joints during sports, especially in football, hockey and rugby. The main causes are a fall on the point of the shoulder, onto an outstretched hand or from a height, like over the handle bars of a bike head first. It can also happen during other high energy injuries, for example, from skiing, slipping on ice, falling off an unprotected height at work or during a motor vehicle accident. These injuries are often called "shoulder separations."

The very outer tip of the shoulder blade (scapula) is called the acromion. The collarbone (clavicle) attaches to the acromion at the acromioclavicular (AC) joint. The acromion serves to protect the ball and socket joint of the shoulder. It supports the deltoid muscle and, through its connection to the collarbone at the AC Joint, is the only direct bone attaching the shoulder joint to the rest of the body. The AC joint can be felt as a prominent bump or ridge at the top of the shoulder. The connection to the collarbone aids in raising the arm overhead and transfers the weight of the arm to the middle of the rib-cage at the sternum.

Pain in the AC joint can result from a specific injury, repetitive trauma, weight lifting or wear and tear. A fall directly onto the shoulder can cause the ligaments surrounding the AC joint to tear. Injuring the AC ligaments, coracoclavicular ligaments, deltoid muscle and trapezius muscle can also result in instability of the AC joint. These injuries range from simple sprains to complete shoulder separations where the clavicle and the acromion are no longer attached to each other.

A "shoulder separation" is a sprain of the AC joint. Shoulder separations are different from shoulder dislocations. A shoulder dislocation involves the much larger ball and socket joint of the shoulder (the gleno-humeral joint), held together by the joint capsule, ligaments and rotator cuff muscles. A shoulder separation involves the smaller AC joint, held together by the joint capsule and strong ligaments around the capsule. The AC joint is further stabilized by strong ligaments holding the clavicle (collar bone) in place, called the coracoclavicular ligaments. In a higher energy accident, the AC joint can dislocate just like the ball and socket of the shoulder. In more severe cases, all of the ligaments holding the clavicle in place are also torn. The strong ligaments holding the clavicle to the coracoid process (another bone in the shoulder) are shown in Appendix Ib.

In a shoulder separation, the injury can range from grades 1 through 6. Grades 1 (a strain of the ligaments with no displacement of the joint) and 2 (displacement less than the width of the collar bone) can be treated with rest, ice and an anti-inflammatory. Grade 3 (displacement of more than 1 width of the collar bone, but a separation less than 2 widths) can also be treated non-operatively, although there is some room here for judgment based on symptoms, arm dominance and activity requirements. Grades 4, 5 and 6 have higher degrees of displacement. The clavicle can be displaced posteriorly (like in a grade 4 injury), displaced more than two widths of the AC joint above the acromion (as in a grade 5 injury) or stuck under the acromion or coracoid (as in a grade 6 injury). In more extreme grade 4 cases, the bone is 'buttonholed' through the trapezius muscle, and there is visible tenting of the skin. This is irreducible and requires surgery as do the other complete displacements of the AC joint (grades 5 and 6).

How do you diagnose an AC joint separation?

The diagnosis of a separation is made by history and a physical examination. The injury causes pain and difficulty with moving the arm. Sometimes it is very obvious on exam, like the grade four separation shown below. Motions behind the back (called internal rotation) will be painful (e.g., washing your back, taking a wallet out of your back pocket or putting on a bra). You may have some swelling, ecchymosis (discoloration of the skin because of blood loss

from a blood vessel underneath it), deformity, tenderness or abnormal translation of the collarbone. Sometimes the skin has an abrasion scab on the skin at the point of contact. An X-ray can confirm the separation..

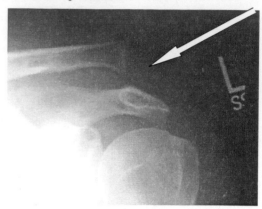

If the ligaments holding the AC joint in place are completely ruptured, the clavicle will move upwards and backwards. Patients may complain of popping, catching or pain with overhead activities. The deformity may be very visible (a prominent bump) and disconcerting, but the deformity itself is not the true indication for surgery. Pain, loss of function, tenting of the skin and instability in the dominant arm are more important factors.

Nonsurgical treatment

A simple dislocation is only a sprain, and the clavicle will not move too much out of place. It can be treated conservatively. Nonsurgical treatment involves NSAIDs, rest, ice and a sling. Certain physical therapy programs will also help. You must stop doing any painful activities or motions.

Surgical treatment

Surgery should not be performed for

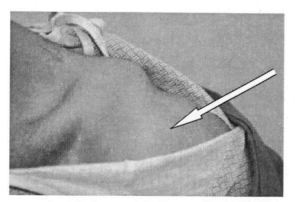

Grade five AC Joint separation.

small separations, minimal deformity or for only cosmetic reasons. When there is a significant deformity, as in grades 4, 5 and 6, surgery is indicated.

The surgical options vary greatly and, until recently, were not as predictable as we would like. Older procedures, like transferring a local ligament from the acromion to the clavicle (Weaver –Dunn procedure) and a screw from the top down (Bosworth screw), have been replaced by newer procedures with better results. When the bump alone is an issue, resection of the distal clavicle may help. Be aware that in some cases, this further destabilizes the bone and increases the movement of the clavicle. Dynamic muscle transfer and various combinations of these procedures have also been used. Most recently, arthroscopic procedures with novel suture anchors, washers and loops have been presented. These may have a roll in acute injury, and they can incorporate a tendon graft. They are technically difficult, and the long term results and complications are not fully analyzed yet.

The current gold standard, in my opinion, is the procedure well described by AD 'Gus' Mazzocca, where the coraco-clavicular ligaments are reconstructed with a tendon graft and locked in place in bone tunnels. In this AC joint reconstruction surgery, a special incision is made over the front and top of the AC joint. The collar bone is reduced into position, and the torn tendons are reconstructed with a tendon graft through tunnels in the clavicle designed to reproduce the anatomy of the torn ligaments. Small, biodegradable screws are placed, supplemented with heavy duty sutures (Fiberwire and Ethibond). The muscle sleeve is then repaired over the now reduced clavicle, reinforcing the ligament repair. It is performed under light, general anesthetic, with, in my practice, local anesthesia containing both Marcaine (a long lasting Novocain-like drug) and Lidocaine.

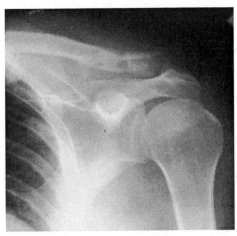

X-ray after reconstruction of the coraco-clavicular ligaments.

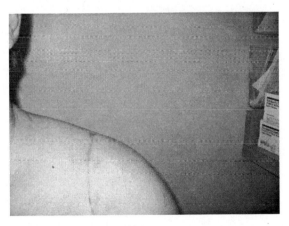

Healed shoulder incision after operative repair of the coraco-clavicular ligaments for AC Joint separation.

Mostly, surgery is very successful. However, there is always a small risk of complications, such as infection, failure of the repair, loss of reduction, coracoclavicular ossification and post-operative arthritis. Having surgery for minor indications is not worth even the limited risk and is discouraged.

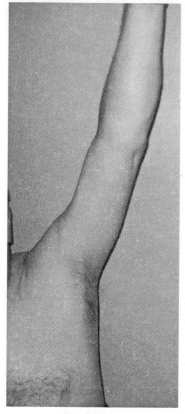

Elevation

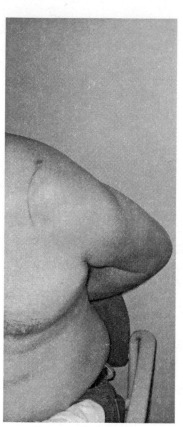

Internal rotation

Four months after repair.

Clavicle Fractures

Believe it or not, the clavicle is the only true bone connection between the entire arm and the rest of your skeleton. You may wonder how this is possible since your shoulder and arm seem to be directly connected to your body. To understand this better, think of how far your arm and shoulder can move. Think of how you can touch the middle of your back, your toes and the back of your head. Try this yourself and feel your shoulder blade move on your back. Feeling its movement, you can tell it is not attached directly to the chest. It rides both up and down and in and out over the ribs, controlled by an amazing group of muscles. Now, feel your collarbone (clavicle) as you move. It is attached to the shoulder blade through a small joint near the top outer tip of the blade, just next to where the deltoid muscle (the largest muscle in your shoulder) attaches. If you follow the collarbone with your finger back towards the center of your body, you will find it attaches to the upper part of your sternum. That joint, the sternal-clavicular joint, is the only point where your whole arm attaches to the rest of the skeleton. It's hard to believe! But, because of this single attachment, the arm and shoulder are allowed a good deal of freedom in their movement. Because the shoulder is attached at one point near the middle of the upper chest, to a bone that's shaped like a thin stick, you can touch almost any part of the rest of your body with little difficulty.

Understanding that the shoulder is attached to the body only through the collarbone helps explain why the clavicle is frequently fractured (broken) when we fall. Its thin, long shape also explains why it fractures with a direct blow, as in a football tackle. Since it is connecting the body and the arm, it can break if the arm is used as protection during a fall. It is also frequently injured during falls off of a bike or motorcycle accidents.

The clavicle typically breaks in the middle. These breaks can be simple (just two pieces, with little displacement) or complex (many pieces and/or displaced). They can be closed (not breaking the skin) or open (breaking through the skin). The clavicle can also break near either end. When it breaks, the acromioclavicular (AC) joint or supporting ligaments could also be involved. The injury can occur in young children, teenagers and active adults. The location of the fracture, the angulation, the displacement, the skin status and the age of the patient are all important in considering treatment.

Treatment for clavicle fractures

In young children, the fracture ends are often close, but sometimes there may be a lot of deformity after a fracture. In growing children, the bones will heal even with a lot of displacement or angulation. In general, many simple fractures can be treated with a sling until the fracture starts to mend (or as we often say, gets sticky). A good sized lump may appear as new bone is made and the fracture starts to heal.

Then some arm movement will be permitted. When X-rays show good healing, the patient can gradually resume activities. In time, and with proper treatment, the lump of new bone that first appears will remodel itself back to the shape of the smooth bone. When the fracture is more complex, tenting the skin or other structures are involved, surgery can be necessary.

Indications for operative treatment of clavicle fractures

In general, clavicle fractures in children are treated in a sling. However, there are exceptions: fractures that break through the skin (open fractures), fractures that threaten to break the skin by the nature of their sharp bone ends and how much they tent the skin, fractures with muscles trapped between the bone ends and fractures causing nerve compression. The same is true in adults. If any of the above issues occur, the fracture should be surgically repaired.

There are a few additional considerations for adult patients. One significant factor is the shortening of the fracture. "Shortening" means that the bone is overlapped more that 2 cm or is 2 cm shorter than normal. In those cases, it should be fixed. Other doctors would add that if it is angulated more than 45 degrees, and if the shortening is in the dominant arm or in a throwing athlete, no more than 1 cm of shortening can be accepted. In any case, patients with deformity, muscle trapped between the fragments or the other factors noted already require an open reduction and internal fixation.

Fractures near the acromioclavicular (AC) joint or the outer end of the clavicle may require special care. The fractures that are lateral to the coraco-clavicular ligaments (discussed in detail in another chapter) seem to heal well if the ligaments are intact. If the ligaments are torn or the fracture is medial to at least one of the two coraco-clavicular alignments, there is a high rate of non-healing (or non-union). These need to be fixed.

If the lateral fragment is small, the challenge is to hold them in place until they heal. Some of these fragments are small and involve the AC joint. When that occurs and the ligaments are torn, removing the fragments and reconstructing the ligaments may be the only viable choice. However, a new type of plate was invented in an effort to solve this complex problem. The plate has a hook designed to catch the acromion. Once the fracture heals, the plate will be removed. The hook plates are not without complications, and excision of the small fragments with a repair or reconstruction of the torn ligaments is also an option.

Fractures of the mid-third of the clavicle do well with non-operative treatment, except for those in which the fracture is displaced posteriorly. That type of fracture can place pressure on major blood vessels and nerves that are just behind the clavicle in this location. In those cases, an open reduction and fixation are needed. The medial-third fractures in children may be growth plate injuries and also need to be fixed.

Surgical Fixation of a Displaced Fracture of the Clavicle:

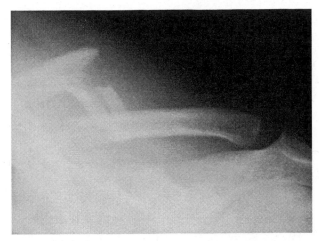

Figure 1: Comminuted fracture of the left clavicle

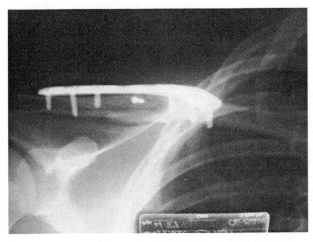

Figure 2: X- Ray of a right clavicle fracture reduced with plate

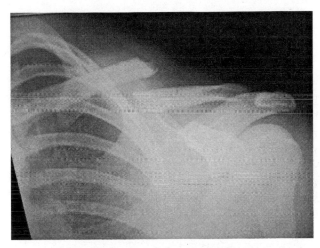

Figure 3: Simple right clavicle fracture with shortening

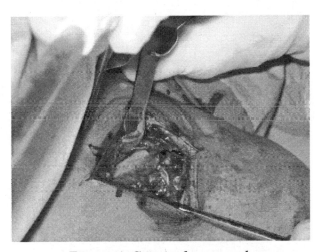

Figure 4: Surgical approach

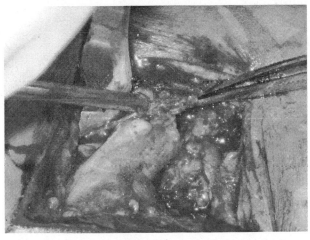

Figure 5: Medial fracture fragment

Figure 6: Lateral fracture fragment

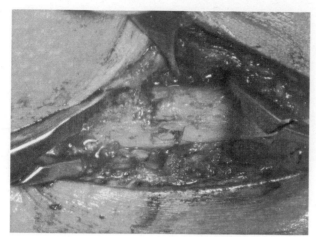
Figure 7: Fracture reduced

Figure 8: Plate in position

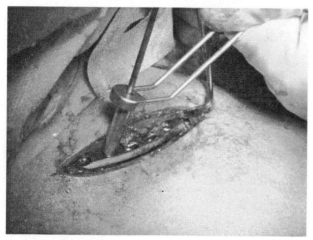
Figure 9: Holes being drilled

Figure 10: Screws being placed

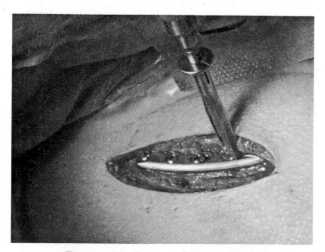
Figure 11: Screws being placed

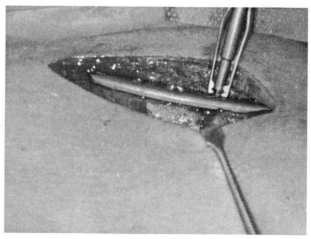
Figure 12: Tightening screws

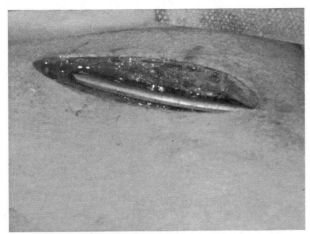

Figure 13: Plate in place with screws

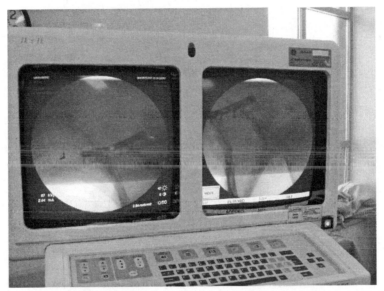

Figure 14: Checking fracture and plate position after reduction with fluoroscopy

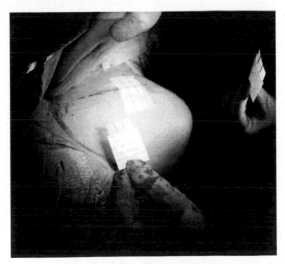

Figure 15: Wound closed

Appendix Ia: Knee Anatomy

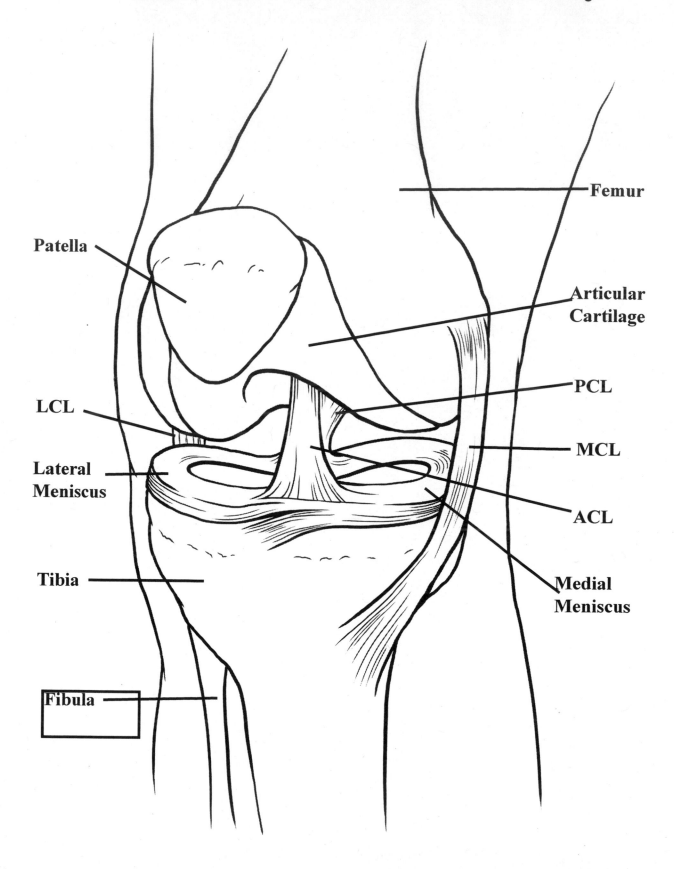

Patella

Femur

Articular Cartilage

PCL

LCL

MCL

Lateral Meniscus

ACL

Tibia

Medial Meniscus

Fibula

Appendix Ib: Shoulder Anatomy

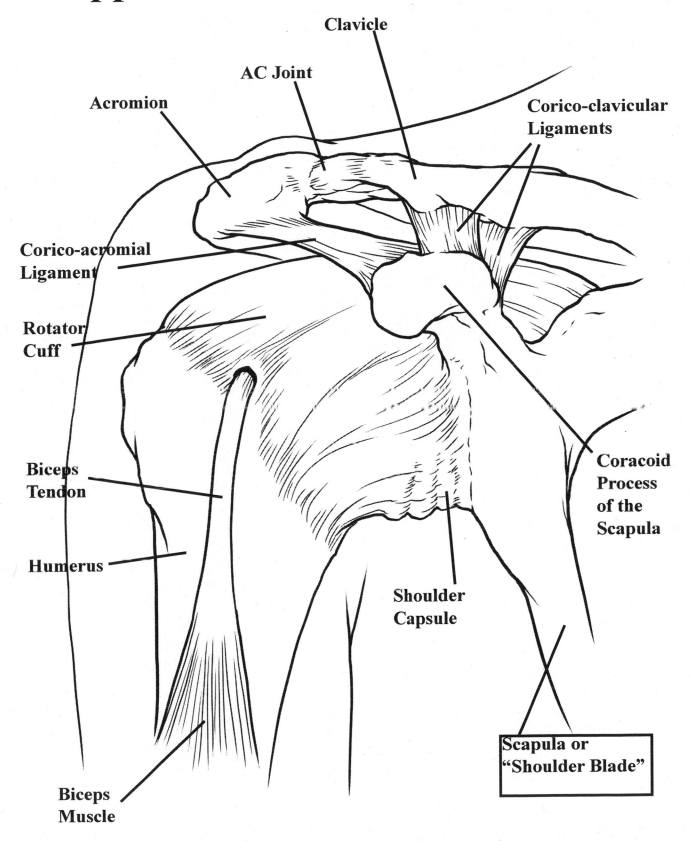

Clavicle

AC Joint

Acromion

Corico-clavicular Ligaments

Corico-acromial Ligament

Rotator Cuff

Coracoid Process of the Scapula

Biceps Tendon

Humerus

Shoulder Capsule

Scapula or "Shoulder Blade"

Biceps Muscle

Appendix II: Physical Therapy

Photos by Elizabeth Reznik

Leg Exercises:

Ankle Pumps: Pump your ankle up and down for one minute as if you were pressing down on the gas pedal. Switch to the other ankle. Do this 10 times per hour while awake. If you're watching TV, do it during every commercial break. This exercise will help decrease swelling, increase circulation, and reduce the risk of developing a blood clot. You cannot overdo ankle pumps.

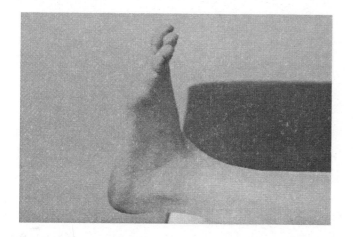 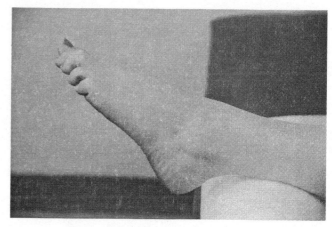

Straight Leg Raises: Tighten your quadriceps (the muscle front of your thigh), and raise your leg 8 to 12 inches off the bed. Alternate legs.

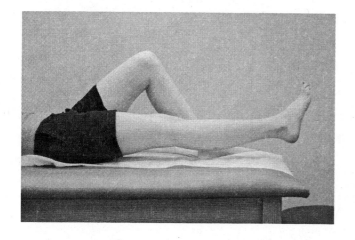 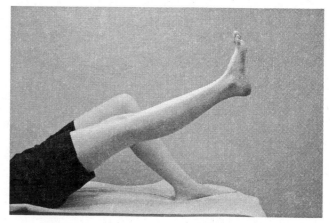

Range of motion: Sit on a chair. Place your foot on the floor [remove your immobilizer and set aside]. Place the uninjured foot under the ankle of the surgical leg. Letting the uninjured side do the work, bring your leg into a straight position. Then, from that position, gently bring the foot down, bending at the knee.

Knee bends/heel slides: With your heel on the bed, bend your knee while sliding your heel toward you.

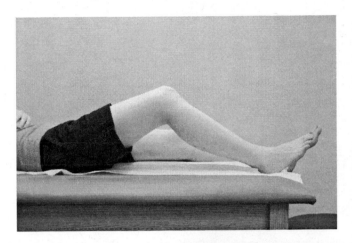 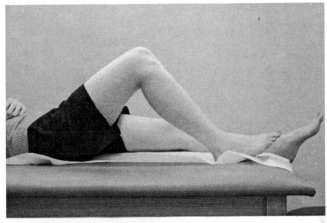

Leg-ups: Lie on your back and flex your hip 90 degrees. Use your hands to stabilize your thigh while kicking upwards. Your hamstrings will act as resistance for your quads exercise.

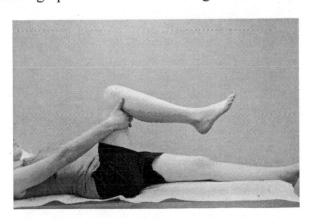 

Arm Exercises:

Hand Squeezes or Grip Strengthening: Using a small soft rubber ball or soft sponge, squeeze your hand. When in the shower, you can use a sponge filled with water. If this is too easy, later in the rehab course you can use a grip strengthener.

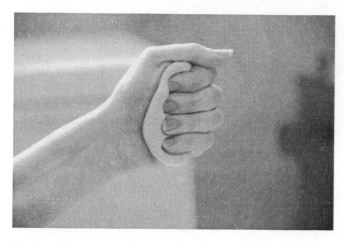 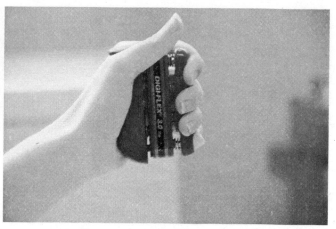

Wrist Range of Motion: Roll your wrist in circles for 30 seconds after each round of grip exercises.

Elbow Range of Motion: Turning your palm inward, towards your stomach, flex and extend the elbow as comfort allows. This will decrease pain and prevent elbow stiffness.

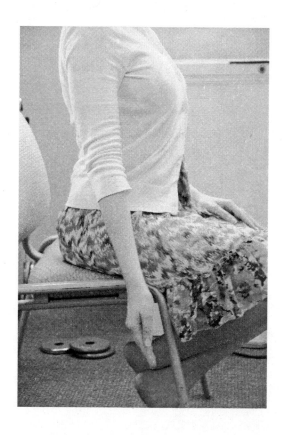 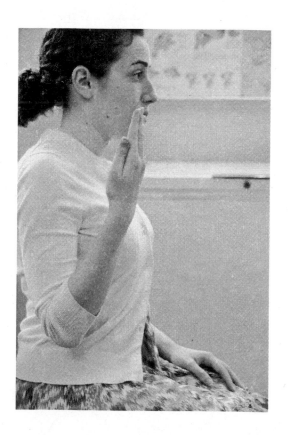

Pendulum Exercise: Holding the side of a table or chair with your good arm, bend over at the waist, and let the affected arm hang down. Swing the arm back and forth like a pendulum. Then swing in small circles and slowly make them larger.

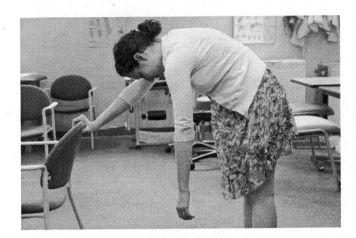 

Wall Walking: Stand facing a blank wall with your feet about 12 inches away. "Walk" the fingers of the affected hand up the wall as high as comfort allows. Mark the spot and try to go higher next time. When more comfortable and stronger (not before three weeks) do these exercise sideways, with the affected side facing the wall. Do not let the hand drop down from the wall- walk your fingers down as well as up. Dropping the arm will strain the repair and be painful. If you feel weak on the way down, feel free to use the other arm to help.

Biceps Curls: Curl the arm up and down 12 times; rest for one minute and repeat for a total of 3 sets of 12. When comfortable, try it holding a very small can. In a few days you can increase can size, but only as comfort allows. This exercise should not be painful. If painful, decrease or eliminate the can.

Wriit Flexion and Extension: Flex and extend your wrist to reduce stiffness and decrease arm swelling.

 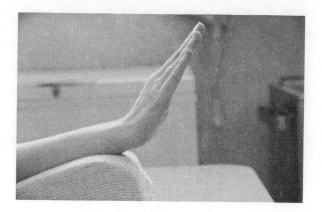

Appendix III: Q and A with Dr. Reznik

How do I know when it's time to see a doctor?

When there is persistent pain with activity, an acute injury with any deformity at all, recurrent swelling of any joint, locking, buckling, giving way or an inability to use the limb.

Is surgery always the best option?

No, many times non-surgical treatments can solve even what would seem to be a complicated problem. A good diagnosis (knowing what is exactly wrong) is very important.

Is surgery painful?

When surgery is needed, it is done with an anesthetic, so there is no pain during surgery. After surgery, Novocain, ice and pain medication should help you throughout the early post-op period and therapy.

Will there still be pain after the surgery?

At first there is always post surgical soreness, stiffness and swelling. Following your doctor's instructions is very important. You cannot outsmart Mother Nature. Elevating the limb means above the heart, using crutches means using crutches. If you cheat on the instructions, you only cheat yourself of the value of your doctor's experience.

What time is my surgery?

The surgical facilities decide the schedule. This is usually done the day before surgery, and surgery times are set up to accommodate many special medical conditions. Therefore, people with medical conditions and small children usually get the earlier operating times.

Will I have to stay overnight?

Most arthroscopic surgeries are done on an outpatient basis. This means you will go home the same day if your procedure can be done arthroscopically (as the majority of procedures in this book are done).

Will I be awake during the surgery?

Most surgeries are done under light general anesthesia along with a local anesthetic to the operative area. (This will help provide comfort after surgery.) So you will not be truly awake during surgery. Some patient's surgeries can be done with sedation and a local anesthetic. Many of these patients will be awake and even talking during the surgery. Alas, many of the anesthetic agents affect short term memory and the patients cannot remember a word of what they said or the procedure itself. Even if you were "awake" for your surgery, you may still be drowsy upon discharge and will need someone to drive you home. You cannot drive yourself under any circumstances.

Do I need to see my primary care physician before surgery?

At many centers, if you are over the age of 40 or have any health problems, you will need to have a "pre-op clearance exam" with your primary doctor. This will include a physical exam and a review of any current medical conditions and medications. EKG, chest X-Ray and blood work will also be done. Patients under age 40 and in good health will only need blood work done. ** Please advise your surgeon of all medical conditions and medications you may be taking prior to surgery. No one wants to be surprised in the OR about something you have been ignoring and hoping will just go away.**

What should I wear/not wear to surgery?

You should wear loose fitting, comfortable clothing that will fit over bulky dressings. Loose, elastic waist shorts or baggy gym type pants are good for knee surgery, and a loose-fitting, button-down type shirt is good if you are having shoulder surgery.
Do not wear jewelry, finger nail polish or contact lenses.

Can I eat or drink anything before surgery?

You may NOT have anything to eat or drink after midnight before the surgery. This even includes water, coffee, tea, Jell-O or even broth. Nothing means nothing.

Should I take my daily medications the morning of surgery?

If your primary care physician decides that you should take your medication in the morning, you may do so with a tiny sip of water. Please check with your doctor at your pre-op clearance exam. This is particularly important for patients with high blood pressure or diabetes. People with high blood pressure should, as a rule, take their medication. Diabetics will not be eating before surgery and typically don't take their medications before surgery, but this is decided on a case by case basis.

When can I shower?

I let my patients remove the dressings and shower after 48 hours with most arthroscopic surgeries. Patients with larger incisions, such as an ACL reconstructions, Fulkerson procedure, AC Joint reconstruction or clavicle repair, will have the dressing removed at their first physical therapy appointment. Those patients can shower after the dressing is changed but need to keep the incision dry.

When can I drive?

In general, you must be off all pain medications and be able to bear full weight on the affected leg if you had knee surgery. If you've had shoulder surgery, you must be completely out of the sling and be able to easily place your hands at the 12:00 position on the steering wheel and move them freely from the 9:00 – 3:00 position.

When do I start physical therapy?

Physical therapy generally starts the 3rd – 5th day after surgery. You should schedule the first several physical therapy appointments at your pre-op appointment with your doctor.

When can I return to work?

It will depend on the type of surgery that you had, as well as the type of work you perform. Your doctor will review your work status at your first post-op visit.

How soon can I return to sports after surgery?

For **meniscal** injuries, it depends if you are having a repair or removal of a torn fragment. For repairs, I usual don't let patients return to full sports for at least 4 months and then only with a brace on - the repair takes at least 4 months to be 80% healed. If the cartilage was removed and not repaired, then 6 weeks is usually okay, but a lot of prep is required for soccer. Remember, soccer, football, rugby and basketball are the toughest sports for your knee after surgery.

For **ACL** tears, this is one of the most commonly asked questions, and no one likes the answer. The ACL is reconstructed using a tendon graft from any number of sources. The key word here is "reconstructed". The tendon is donated to the site, and, once harvested, the graft tissue has no blood supply. Therefore, the body has to heal the tendon to the bone and then grow new blood vessels into the new ligament. The bone may heal to the tendon in 2-3 months, but the creation of new blood vessels takes much longer. In fact, the tendon graft gets softer as this occurs, and you are at a higher risk of re-injury between 3-6 months after the surgery. They can start with a fast walk at 6 weeks, light jog and run straight, with no cutting and on a soft track, after that. I let my patients bike, use an elliptical trainer, tread mill and swim (with no whip kick) after 3 months. Some can go back to light sports with a brace on after 6 months. In general, ACL patients must wait 9 months or more for heavy sports (soccer, basketball, football, etc.). At nine months, they can start sports specific training, and at one year, they can usually return to a full competitive or contact level. This sounds long, but cheating this timeline creates a high risk of re-injury and the need for a second surgery.

How can I prevent re-injury?

For ACL tears, this is probably the second biggest concern for patients: the fear of re-injury.

To prevent this, many of my patients return to play with an ACL brace on, even after the ACL is reconstructed with a good clinical result. Many will give up the brace in time, but some need the extra comfort and use a custom ACL sports brace for all sports and heavy activity. These patients with poor jumping stlye (valgus collapse, see section on ACL tears) or poor hip control/balance need to work on these skills before returning to any sport.

Is it better to repair or cut out the torn part of my meniscus?

It is better to save the parts of the meniscus that can be. In general, it is best to repair the menisci of younger patients with tears in the zone with the best blood supply. There is a balance between saving the meniscus and the success of the repair. If a repair has a 90% chance of failing, it should not be done and visa-versa; if there is a good chance the tear will heal, it should be repaired.

Will my meniscus ever grow back?

The meniscus does not truly grow back, some scar will form at times, but it never has the weight bearing features of the original meniscus.

When can I drive again?

For knee patients: You need to be off of all pain medications and able to fully weight bear (no crutches) with no pain and step on the break without hesitation. *For shoulder patients:* You must be ready to be out of your sling, off of all pain medications and able to move the steering wheel with both hands fully without pain.

Appendix IV: Making the Most of Your Office Visit

By Denise M Sutcliffe, LPN

We all know that a visit to the doctor is not always a stress-free experience. Doctors are busier than ever these days and it sometimes takes a while just to get an appointment. Patients are also much busier and have less time to spend waiting to see the doctor. Then, once you arrive for your appointment, often you will find the waiting room full, and there are usually several forms you will need to complete, including a patient medical history form and insurance information.

Ok, Relax!

As a nurse working in a doctor's office, I have found that patients who prepare for their visits seem *to have a better overall experience.* Advanced planning can save you time and make your visit less stressful.

Here are some tips to make your visit less stressful **and** more productive:

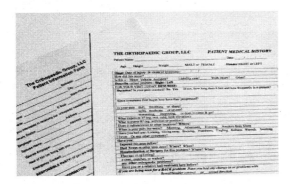

• When making your first appointment with the doctor, ask if the new patient forms can be faxed or mailed to you so you can complete

them at home prior to your visit. For your convenience, these forms can also be downloaded from our website, http://togct.com/your_visit.html.

• You will be asked to complete a new medical history form, even if you are not a new patient, if you haven't been seen in over one year or if we are seeing you for a new problem. Many patients find completing these forms an inconvenience, but they play an important part of your visit.

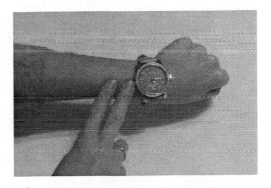

• If you are not able to do the forms ahead of time, try to arrive early for your appointment to complete the paperwork without feeling rushed.

• If you find the waiting room full when you arrive, don't be afraid to ask if the doctor is running behind or how long you may be expected to wait. The doctor may have had an unexpected emergency or several complicated patients that may have put him behind schedule. We understand that you may be on a time constraint and will do our best to adhere to the schedule.

We all know that a visit to the doctor is not always a stress-free experience. Doctors are busier than ever these days and it sometimes takes a while just to get an appointment. Patients are also much busier and have less time to spend waiting to see the doctor. Then, once you arrive for your appointment, often you will find the waiting room full, and there are usually several forms you will need to complete, including a patient medical history form and insurance information.

Ok, Relax!

As a nurse working in a doctor's office, I have found that patients who prepare for their visits seem *to have a better overall experience.* Advanced planning can save you time and make your visit less stressful.

Here are some tips to make your visit less stressful **and** more productive:

• When making your first appointment with the doctor, ask if the new patient forms can be faxed or mailed to you so you can complete them at home prior to your visit. For your convenience, these forms can also be downloaded from our website, http://togct.com/your_visit.html.

• You will be asked to complete a new medical history form, even if you are not a new patient, if you haven't been seen in over one year or if we are seeing you for a new problem. Many patients find completing these forms an

inconvenience, but they play an important part

Inaccurate medication list

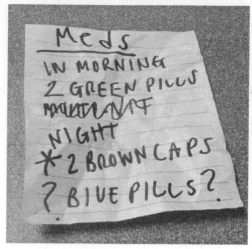

Accurate medication list

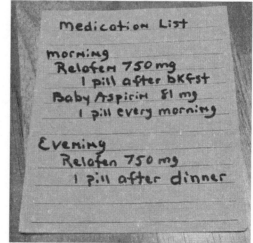

of your visit.

• If you are not able to do the forms ahead of time, try to arrive early for your appointment to complete the paperwork without feeling rushed.

• If you find the waiting room full when you arrive, don't be afraid to ask if the doctor is running behind or how long you may be expected to wait. The doctor may have had an unexpected emergency or several complicated patients that may have put him behind schedule. We understand that you may be on a time constraint and will do our best to adhere to the doctor's schedule.

• Be sure to fill out the medical history form completely. Some of the questions might seem irrelevant to you, such as "are you right or left handed?", "what is your height and *weight*?" (Many patients seem to leave this one blank!) and "what is your occupation and job duties?" However, all this information is important when the doctor may need to determine things like your ability to return to work/school, appropriate medication dosages, etc.

• List any and all telephone numbers where you can be reached, as well as an emergency contact number. It is important that we have a way to reach you in case we have a question or concern.

• Be sure to note if you have any allergies. This is extremely important if the doctor needs to order any medications for you.

• Know what medications you are currently taking. Saying that you take "something for blood pressure" or "2 small red pills" does

not provide the doctor with enough information.

• Have an accurate list of all medications you are taking, including dosing information and how frequently you are taking them. Include all over the counter medications as well. The doctor and nurse need this information

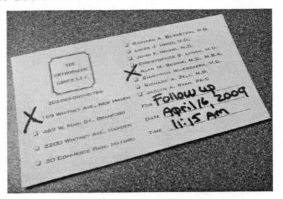

when prescribing any new medications in order to avoid any interactions. It is helpful to list the name and telephone number of your pharmacy in case we need to call or fax any prescriptions for you.

• What are your symptoms? Include when your symptoms began and what you were doing at the time. Be as specific as you can. Saying that you have had pain "for a long time" can mean several days to one patient and several years to another. Have you ever felt these symptoms before? What makes it better? What makes it worse? Is it better or worse after working, waking up? Have you ever had surgery on the affected body part in the past? All this information gives the doctor important clues to help make a diagnosis.

• Have you had any X-rays or MRI films done? If so, be sure to bring them for the doctor to review. This seems to be the most frequent source of frustration for our patients. They are often told that "the films will be sent to the office" and when they arrive, there is nothing here for the doctor to review. Most doctors' offices and Diagnostic Testing Centers **DO NOT** forward X-rays or MRI films/discs. You do need to request a copy and take it with you when you leave the imaging facility. If you were told the films would be sent to the office, it is worth a telephone call to our office to verify that they are here prior to your visit.

• If this is a follow up visit, let the doctor know if your symptoms improved, worsened or stayed the same. If medications were given to you, did they help? Did you take them as ordered? Were any side effects noted? If physical therapy was ordered, did it help?

• During your visit, don't be afraid to ask questions until you understand. If your doctor suggests a test or a change in your treatment, ask why it's necessary and what the results will tell you. When does the doctor want to see you again?

• If a follow up visit is recommended, be sure to schedule it prior to leaving the office. This will allow you to book the appointment time and place that is the most convenient for you.

Denise is a Licensed Practical Nurse with 15 years of experience working as a staff nurse at the Hospital of St. Raphael and as a charge nurse in an area skilled nursing facility. She has been with The Orthopaedic Group, LLC since March, 2006 and works as an Orthopaedic Nurse Associate with Dr. Alan M. Reznik.

She strives to provide excellent and effective patient care on a daily basis.

Glossary

A

Achilles tendon – The strongest tendon in the body, it connects the calf muscles to the heal bone.

ACL (Anterior Cruciate Ligament) – This is a ligament that connects the femur to the tibia. It helps stabilize the knee and stops the tibia from extending past the femur.

Acromioclavicular (AC) joint – This joint, made up of strong ligaments, connects the acromion to the clavicle.

Acromion – The topmost point of the scapula, supports the deltoid muscle and connects the shoulder blade to the clavical.

Acromioplasty – This is a procedure to cure Impingement Syndrome in which damaged tissue of the shoulder can be removed, the sub-acromial space can be increased, and the inflamed bursa can be cleared.

Adhesive Capsulitis ("Frozen Shoulder") – This is a condition in which the capsule becomes inflamed and the joint stiffens. There is pain and a progressive loss of shoulder motion. This is because adhesions (bands of tissue) will develop between tendons and ligaments, restricting bone motion.

Allograft (of cartilage or ligaments) – Transplanting frozen cartilage or ligaments from a donor.

Ambulation – This means walking.

Anesthesia – A drug that causes a patient to lose sensation or consciousness so that surgery can be performed without any pain. It can be "local", in which case it only numbs the region being operated on. It can be "regional", meaning it numbs everything below the waist. Finally, it can be "general", which will put the patient to sleep.

Arthritis – A disease in which the joints become inflamed, causing pain and stiffness.

Arthrogram – Injecting a radioactive substance in order to take a radiograph of a joint.

Arthroscopy – A minimally invasive surgery in which one looks inside a joint with a fiber-optic scope and can perform procedures.

Articular cartilage – See Cartilage

Autograft (of cartilage or ligaments) – Transplanting cartilage or ligaments from one part of the patient's body to another.

B

Biceps – This is a muscle that starts at the elbow and splits into two tendons. The shorter tendon (short head) ends at the coracoid process of the scapula. The larger tendon (long head) enters the shoulder joint.

Biceps tenodesis – This is a surgical procedure in which the ruptured end of the biceps tendon is anchored to the upper end of the humerus.

Blood clot – This is a mass of gelled blood cells.

Bone spurs – Abnormal bone projections that grow off of joints.

Buckling (of the knee) – The knee folds or "gives way" causing you to temporarily lose control of your leg.

Bursa sac – A small, fluid-filled sac made of tissue. They create surfaces for smooth motion.

C

Calcified tendonitis – These are partial tears of the rotator cuff that are associated with calcium deposits in the substance of the tendon.

Capsular Shift procedure – This is done when the ligaments and capsule lining are stretched out of shape. The loose capsule is tightened at the same time as the ligaments are repaired.

Cartilage – A type of connective tissue that covers joint surfaces, cushions the bones, and absorbs shock.

Clavicle – The collarbone.

Coracoacromial ligament – This is a ligament that connects the coracoid process to the acromion.

Coracoid process – This is a part of the scapula. It helps stabilize the shoulder joint.

D

Degenerative tears – These usually occur when cartilage weakens and thins over time. It becomes more prone to tear from simple motion.

Dislocation – When a joint slips all the way out of place.

Distal – This describes something that is situated away from the center of the body, like limb.

F

Femoral condyles – These are the two large bone projections that stem from the bottom of the femur. The lateral femoral condyle is towards the outside of the body and the medial femoral condyle is towards the inside.

Femoral notch – The roof of the center of the knee, between the two femoral condyles.

Femur – The thighbone.

Fracture – A bone is cracked.

G

Glenoid cavity – This is the shoulder socket.

Glenoid labrum – This is the rim of tissues surrounding the glenoid cavity.

Gout – A disease that causes inflammation of

the joints and excessive uric acid in the blood stream.

"Greenstick" fractures – Fractures that occur in the middle of children's bones since they can bend without breaking.

Growth plates – Softer, developing bone in growing children. They cannot be seen in X-rays.

H

Hamstrings – These are muscles behind the thigh that help people run.

Hill Sach's Lesion – This is damage to the head of the humerus.

Humerus – The upper arm bone. The head of the humerus rests in the glenoid.

Hyperextension – Extending a joint beyond its normal range.

Hypertension – High blood pressure.

I

Impingement Syndrome – This is a condition that results from overuse of the shoulder. The tendons and bursa may thicken and pinch against the bone, causing irritation and pain. It can also occur at the AC joint.

Insidious – In medical terminology, insidious means that a disease develops so gradually, that it is usually well established before it is diagnosed.

L

Labrum – The cartilaginous lip that covers the edge of the glenoid cavity.

Lateral – This describes a body part that is located away from the center of the body. It could be used to refer to the side of a body part that is towards the outer side of the body.

Ligament – Tough tissue connecting bones to bones.

Locking (of the knee) – The knee is frozen in a bent or extended position; it is unable to be fully extended or flexed.

Loose bodies – These are cartilage fragments that detach and are loose in the knee.

M

Medial – This describes a body part that is located towards the center of the body. It could be used to refer to the side of a body part that is towards the inner side of the body.

Meniscal repair – This is a procedure in which the torn part of the meniscus is repaired.

Meniscus – Cartilage between the femur and tibia. It cushions the knee and distributes the weight from the femur onto the tibia. It helps with stability of the knee joint by converting the flat top of the tibia to a more stable, shallow socket.

Microfracture technique – The bone surface is

drilled in order to help blood and marrow get to the surface.

MRI (Magnetic Resonance Imaging) – Strong magnets and radio waves are used to create images of the soft tissues in the body. The pictures are clearer than X-rays.

Mumford Procedure – This is resection of the prominent tip of the distal clavicle.

N

NSAIDS – This stands for Non-Steroidal Anti-Inflammatory Drugs (for example, aspirin and ibuprofen).

O

Osteoarthritis – A disease in which the cartilage of joints breaks down over time.

Osteochondritis Dissecans (OCD) – When fragments of bone below a joint lose blood and separate from the rest of the bone.

P

Partial meniscectomy – This is a procedure in which the torn part of the meniscus is removed.

Patella – The kneecap.

Patella Tendon – A large tendon connecting the thigh (quads) muscle and patella to the tibia. It holds the patella in place.

Patellofemoral Articulation – This is the name for the smooth surfaces between patella and trochlea groove.

Periosteum – This is membrane that lines the outside of all bones, except for the end of long joints.

Proximal – This describes something that is situated close to the center of the body.

Q

Quadriceps – The thigh muscle of the leg. It controls the movement of the patella.

R

Recurvatum – Hyperextension of the knee.

Red zone (of the meniscus) – This is the outer third of the meniscus; there is blood flow.

RICE – This stands for Rest, Ice, Compression and Elevation.

Rotator cuff – This is made up of four muscles and their tendons that originate from the scapula and form a single tendon unit that inserts on the upper humerus. It helps stabilize the shoulder within the joint, lift the arm, and rotate the humerus.

S

Scapula – The shoulder blade.

Scar tissue – This is connective tissue forming a scar. It is composed mostly of fibroblasts and unorganized collagen

Sprain – The stretching or tearing of a ligament.

Subluxation – Popping out of place, a partial dislocation.

Synovitis – An inflammation of the lining in the knee or shoulder. It can also occur in the elbow, wrist, or ankle.

T

Tendon – Tough tissue connecting muscles to bones.

Tibia – The shinbone.

Tibial tubercle – This is the bony prominence on the tibia below the patella. The patella tendon attaches here.

Trapezius muscle – We have two of these muscles, each located on either side of our upper backs. They are mainly used to stabilize and rotate the scapula.

Trochlea – This is the groove in which the patella glides up and down.

Tuberosity – A prominence on a bone where ligaments or tendons, like the rotator cuff will attach. The humeral head has greater and lesser tuberosities. The intraspinatus and supraspinatus attaches to the greater, and the subscapularis attaches to the lesser.

V

Valgus alignment – The legs bow outward.

Varus alignment – The legs bow inward.

W

White zone (of the meniscus) – This is the inner third of the meniscus; there is no blood flow here and therefore very little potential for healing.

X

X-ray – Electromagnetic radiation that bounces off bones to create a photograph of part of a person's skeleton. It is used to make diagnoses.

INDEX

Author's Biography

Alan M. Reznik is board certified in Orthopedic and Arthroscopic surgery, specializing in sports medicine. A Westinghouse Science talent search honoree, he received his Bachelors of Science from Columbia University's School of Engineering. There, as a member of a research group studying Sudden Infant Death Syndrome, he became interested in medicine. Later, his expertise in engineering merged with his love of medicine while attending Yale University School of Medicine. He pursued a specialty in orthopaedic surgery. During his residency at Mount Sinai Medical Center in NY, he served as a court physician at the US Open Tennis Tournament for four years and was selected for a fellowship at Oxford University. He then completed a fellowship in sports medicine at the University of California, San Diego with Dr. Dale Daniel, a world-renowned knee ligament expert, and Dr. Raymond Sachs, a ground breaking shoulder expert.

In New Haven, Connecticut, Dr. Reznik was a founding member of the Yale-New Haven Hospital's Orthopaedic Trauma team and awarded the Yale Residents' Teaching Award. He served on the game organizing committee for the 1995 Special Olympics, where he helped care for special athletes from over 105 countries. He was also chosen to be the team physician for the New Haven Knights professional hockey team. Dr. Reznik has also served as a volunteer surgeon on a medical mission to New Orleans in the aftermath of Hurricanes Katrina and Rita in 2005. The team visited many sites and set up a clinic in a domed stadium to care for displaced residents of New Orleans living there without health care. In 2008, he visited Cuba and Bolivia, Cuba on a humanitarian mission and Bolivia to see, first-hand, the role of "Save the Children" in the poorest-of-poor countries in South America.

Dr. Reznik was selected by Connecticut Magazine as one of the "Top Docs" in the state by patients, nurses, physicians from other specialties and his orthopedic surgeon peers for over ten years. A member of the Arthroscopy Association of North America since 2001, he has also been named as one of "America's Top Physicians" by the Consumer's Research Council of America on many occasions over the last several years and again in 2009.

Dr. Reznik has invented several instruments for knee and shoulder arthroscopy. He holds a patent on his shoulder holder, an arthroscopic fluid control system and has a number of additional patents pending. His inventions are designed to improve the effectiveness of arthroscopic surgery. Currently, his arthroscopic suture grasper is being marketed by Johnson and Johnson, and his arm positioning device is being manufactured by Innovative Medical Products. These products were introduced at the 2009 Convention of the American Academy of Orthopedic Surgeons.

In an academic private practice for twenty years, Dr. Reznik is the managing partner of

The Orthopaedic Group, LLC in New Haven, CT. He serves on the Board of Directors at Ezra Academy. He enjoys golf. He writes articles for the group's website, monthly newsletter and patient magazine. Each year, he writes his own holiday greetings, and his first novel will be available online later this year. In his daily practice of medicine, Dr. Reznik enjoys caring for pediatric and adult recreational, competitive, professional and 'working' athletes each and everyday.

Jane Y. Reznik is currently a student at the University of Pennsylvania majoring in Biology. Before starting at the University of Pennsylvania she graduated with honors and was elected to the Cum Laude Society by the Hopkins School in New Haven Connecticut. At Hopkins, she excelled in the sciences, was a peer leader, peer tutor and was recognized with the Josiah Willard Gibbs award for science in her senor year. She worked for the "Little Scientists," a local education company that has made its reputation in bringing science and science learning to children of all ages. There, she edited and helped organize science materials for their varied programs. Jane has carried these unique skills forward and applied them to her work on this book.

Jane has distinguished herself by volunteering her time teaching those who are less advantaged in Philadelphia. In 2008, she traveled to Bolivia to work with an education program run by Save the Children and continues to tutor children in mathematics regularly.

Jane is currently working at the Yale University School of Medicine's Pulmonary Laboratory under Dr. Jack Elias and Dr. Charles Dela Cruz. Her work is related to gene pathways activated by cigarette smoke. Jane loves science and its application to medicine.